衣者・天地形

Clothing : the Reflection of the Origin

衣者·天地形

Clothing : the Reflection of the Origin

李薇 著
Li Wei's Works

中国纺织出版社有限公司

序

立足神妙 着力微妙

刘巨德

2004年，李薇的艺术作品《夜与昼》在全国美展荣获金奖，一下轰动美术界，尤其震撼服装设计界并引发当代艺术圈关注。为什么这件作品有如此大的魅力？我想在于它的原创性，关键这原创行为模糊了设计艺术与纯艺术的边界，在看不见的艺术深处继承了传统，这是艺术界少有的，非常值得我们大家品读。

从气象上看，整件作品犹如凤凰的幽灵飞翔夜与昼，飘游于透明的蝉翼般的丝织品经纬网络间，释放着东方神秘主义的光芒，完全超越了服装艺术形而下的实用性，走向形而上的纯美大美境界。气象宏大，微妙空灵，如诗如歌，如梦如幻，或像一个夜与昼的神话。从远古来，却没有远古传统的影子，只有永恒的星光闪烁在纱幔里，长风、皓月、流水、黎明、曙光，轻歌曼舞，在夜与昼中弥漫无尽。

再细观作品的制作，丝料变形了，垂直水平的经纬线错位了，有序变为无序……李薇为她的艺术，把尚好的丝织品撕裂了，她说自己"我很坏"，我理解她的自语，艺术家都是挑战规则的人。李薇大幅度的破坏和颠覆，创造了从破坏到重建的奇迹，即旧的破坏了、新的诞生了，这是庄子"凡物成与毁，复通为一"的天地创造之大道。

李薇不凡的手，从细微的丝织品经纬线上，摸到了这大道的玄妙，至使她的《夜与昼》，出现了不可预见的不确定性的神奇，无序又有序，细微又宏大，可能这是服装艺术界没有的。《夜与昼》前无范例，后无参照，李薇将艺术创作中限制与自由的矛盾化解。丝织品正常静止的方格经纬线，在她手里化为了伸向天宇的不规则的运动曲线，这曲线不断放大，层层叠叠，无序变为有序，穿裹在半裸的女人体上，瞬间人与自然之道相合，艺术魔力不期而遇。

《夜与昼》证明纯艺术与设计艺术没有边界，相互没有限制，设计艺术家本质上都是纯艺术家，边界只存在于形而下层面，在形而上层面无界。真正的艺术家逍遥于行而上和行而下之间，逍遥而无分别，致大致远，这是我院老一辈艺术家治学的传统，他们身上都长着设计艺术和纯艺术两只翅膀，李薇在看不见的艺术深处与他们的灵魂相遇。

如果说，设计艺术是鱼的话，纯艺术就是水，真正的设计艺术家既是水也是鱼。他们心怀大海，其鱼化而为鸟，始生大鹏，徙南冥，化为凤凰。李薇的《夜与昼》正是凤凰落于她手上的结果。我祝贺她，这是境界，也是她人格的力量所至。她谦和、温厚、慈祥、智慧的样子总与她的微笑同在。

2023年9月17日于荷清苑

Preface

Grand Vision, Subtle Details
Liu Jude

In 2004, Li Wei's art piece "Night and Day" won the Gold Award in the National Art Exhibition, causing a sensation in the art circle, especially in the field of fashion design, and attracting attention from the contemporary art community. Why does this artwork possess such great charm? I believe it lies in its originality, as this original creation blurs the boundaries between design art and pure art. Furthermore, it inherits tradition from the invisible depths of art, which is rare in the art world and deserves our careful contemplation.

From an artistic perspective, the entire artwork resembles the phantom of a phoenix, weaving through the transparent and delicate silk fabric, emanating the radiance of Oriental mysticism. It indeed transcends the utilitarian nature of fashion art, venturing into the realm of pure beauty and grandeur. It possesses a grand and ethereal presence, like poetry and song, like a dream or an illusion, akin to a myth of night and day. Originating from ancient times, it carries no trace of ancient traditions, only the eternal starlight twinkling within the veil. The gentle breeze, the moonlight, the flowing water, the dawn, the first light,they all sing and dance gracefully, diffusing endlessly within the realm of night and day.

Taking a closer look at the creation of the artwork, we see the deformed silk fabric with the vertical and horizontal warp and weft lines becoming misaligned, and the order turning into disorder... Li Wei, for the sake of her art, tears apart the pristine silk fabric. She refers to herself as "being bad" and I understand this as her self-expression. Artists are often challengers of rules. Through Li Wei's extensive destruction and subversion, she creates a miracle of destruction and reconstruction, where the old is destroyed and the new is born. This reflects the Taoist concept of "everything returns to unity through destruction and rebirth", which embodies the grand path of creation in the realm of heaven and earth.

Li Wei's extraordinary hands touch the profound essence of this path, even on the subtle warp and weft lines of the silk fabric. As for her work "Night and Day", it possesses an unforeseeable and wondrous element of uncertainty. It is both chaotic and orderly, delicate and grand, surpassing anything seen in the world of fashion art. Without precedent or reference, Li Wei resolves the contradiction between limitations and freedom in artistic creation. The normally stationary grid of warp and weft lines on the silk fabric transforms in her hands into irregular curves reaching towards the heavens. These curves continue to expand, layer upon layer, transforming disorder into order. When draped over the semi-nude female, there is an instant and natural harmony between humans and nature, encountering the enchantment of artistic magic unexpectedly.

"Night and Day" demonstrates that there are no boundaries between pure art and design art, as they do not restrict each other. Design artists, fundamentally, are pure artists, and boundaries only exist on the superficial level while it is ultimately boundless on the metaphysical level. True artists roam freely between the higher and lower realms, transcending distinctions and reaching far. This is a tradition of learing left behind by the older generation of artists in our institute, who possessed both the wings of design art and pure art, and Li Wei encounters their souls deep within the realm of art.

If design art is likened to a fish, then pure art is the water. A true design artist is both the water and the fish. They carry the vastness of the ocean in their hearts, transforming from fish into birds, resembling the great roc, soaring to the southern horizon, and transforming into phoenixes. Li Wei's "Night and Day" is precisely the phoenix that has landed in her hands. I congratulate her on this achievement, as it represents not only a realm of artistic expression but also the power of her personality. Her modest, gentle, wise and kind demeanor always accompanies her smile.

September 17, 2023
Heqing Garden

目录 / Contents

008	学术评价	Academic Comments
030	李薇简介	Li Wei Introduction
036	夜与昼	Night and Day
050	七彩云南	Colorful Yunnan
068	青绿山水	Turquois Landscape
078	新中式	New Chinese Style
094	蓝之韵	Rhyme of Blue
100	红色记忆	Red Memory
112	紫气东来	The Purple Spirit
120	锦瑟华年	Blooming Times
140	荷塘月色	Moonlight over the Lotus Pond
146	雅韵	Elegance
158	花梦敦煌	Flower Dream of Dunhuang
166	空与影	Emptiness and Shadow
180	述说	Narration
184	清·远·静	Qing Yuan Jing
188	行云流水	Flowing Clouds, Running Water
190	风声	Sound of the Wind
194	夜与昼·丝网版画	Night and Day: Silkscreen Prints
198	山峦	Mountain Range
200	枯山水	Dry Landscape
202	大漠孤烟	Smoke in the Desert
204	乐山水	Joyful Landscape
208	大事记	Chronicle of Events

学术评价

国内学术评价

岛子

策展人、清华大学美术学院教授

李薇，其人其艺均淡泊静笃，清逸而温婉。她的艺术实践虽以服装、染织为媒介，但由于其艺术观念所秉有的实验精神，使之从一个优秀的设计师从容蜕化为一个具有相当影响力的优秀艺术家。当然，在艺术史上这种二重性的划分从来都不是绝对的，关键是作为一个当代艺术家，除了需要创新形式的视觉知识水平，同时也要求新时代的艺术教育和观众跨越艺术类别和主题之间的边界，完成审美、伦理和信仰的飞跃，从艺术作品所表征的多种视角，包括流行文化、科学、宗教、历史、政治，来考察和思考现实存在。

李薇教授的艺术法度源自意象美学之气韵审美，多年来着力在服饰染织上实现气韵之风神。"气韵"之说，始自六朝南齐人物画家谢赫的《古画品录》，六法之首的"气韵生动"成为水墨艺术根本性审美标准。谢赫是人物画家，人物画之"韵"来自所谓魏晋风度，也即当时的人伦鉴赏标准，指一个人的情调、个性，有清远、通达、放旷之美，而这种美流注于人的形相之间，从形相中可以看得出来的，把这种神形相融的韵，在绘画上表现出来，即是"气韵"的"韵"。六朝伊始，中国人"物感"意识开始觉醒，所谓"物感"，即人与物的交流互感，人从自然万物中感受到了自然宇宙生命的活泼、真实，并从而赋予大自然活泼、真实的情感特质，所谓登山观海，情满意溢，正如南朝画家王微在《叙画》山水画论文中说山水有情，"望秋云，神飞扬；临春风，思浩荡"，意即仰观秋云，临风而立，人与大自然的感情相互印证，在人与物交流互感中，体会一种无言之美。

李薇的气韵审美在时装秀场中展现得淋漓尽致，通过模特的身体、表演的空间、现场氛围、交互感以及出人意料的并置、连结、绵延、衍义等，呈现出飘逸、放达的出世情怀和现代人神往的自由态势，她的作品和秀场将动与静、虚与实、悠远与亲近、视觉性与写意性、古典与前卫相互融合、辩证、转化，相互塑造、启动，而其大幅度的、连续性的壁挂、屏风作品充当着秀场的阐释透镜，精妙、鲜活地阐释出水墨山水的林泉韵致，人在山水境界的忘我、超逸，因而令人身临其境，神游其间，流连忘返。

需要提示的一点是，身体性既是本展览的隐性主题，也是李薇从学院艺术向实验艺术转型的一个向度。身体或身体性作为艺术场域和伦理场域，大于一般艺术媒介，因此它可以和法国哲学家吉尔·德勒兹的块茎(rhizome)概念通用，"块茎"这一术语描绘互相联结的研究和思想，它们不但无起点和终点，无贯穿系统内部的固定通路，还拒绝严密僵化的组织形式和支配性概念，并且有能力将异质性元素连接在一起。在此意义上，身体或身体性作为当代艺术的"块茎"，就在于当今许多视觉表征和阐释的知识体系都是流动的、非等级的、非线性而是去中心化的。身体

Academic Comments

Domestic Academic Comments

Daozi

Curator, Professor of Academy of Arts and Design, Tsinghua University

Li Wei is a person of both talent and virtue, modest and serene, delicate and graceful. Although her artistic practice is mainly focused on clothing, dyeing and weaving, her experimental spirit, derived from her artistic concept, has transformed her from an outstanding designer into a highly influential artist. Of course, in art history, such a dualistic identification is never absolute. The key point is that as a contemporary artist, one needs a new level of visual knowledge and requires art education and audiences in the new era to transcend the boundaries of artistic categories and themes. This is to accomplish a leap in aesthetics, ethics, and beliefs, by examining and contemplating reality through diverse perspectives represented in artworks, including popular culture, science, religion, history, and politics.

Professor Li Wei's artistic principles are derived from the aesthetic of imagery and the beauty of rhythm. Over the years, she has focused on achieving a vibrant and spirited style in clothing dyeing and weaving. The concept of "Rhythmic Charm" originates from Xie He, a figure painter from the Southern Qi dynasty during the Six Dynasties period, and his book "Record of Famous Paintings of Ancient Times," in which "rhythmic vitality" is listed as the first of the Six Canons, serving as the fundamental aesthetic standard in ink painting. Xie He is a figure painter, and the "rhythm" in figure painting originates from the so-called Wei-Jin style, which was the aesthetic standard of human relations at that time, referring to a person's temperament and personality, embodying a sense of purity, openness, and unrestrained beauty. This beauty flows between the physical appearance of a person and can be observed from their appearance. The expression of this harmonious fusion of spirit and form in painting is referred to as the "rhythm" of "Rhythmic Charm". At the beginning of the Six Dynasties period, the Chinese people's consciousness of "sensibility towards objects" began to awaken. The so-called "sensibility towards objects" refers to the interaction and mutual influence between humans and objects. People sense the liveliness and authenticity of natural life from the myriad of natural phenomena, and in turn, they imbue nature with lively and authentic emotions. It is like the experience of climbing mountains and observing the sea, where emotions overflow. Wang Wei, a painter from the Southern Dynasty, stated in his essay on landscape painting in "Xu Hua" that landscape is emotional. He wrote, "Beholding autumn clouds, my spirit soars; Facing the spring breeze, my thoughts expand boundlessly." This means that when gazing at autumn clouds and standing in the wind, the emotions of humans and nature mutually reflect and confirm each other. Through the exchange and mutual influence between humans and objects, one can experience a wordless beauty.

Li Wei's aesthetic of rhythmic charm operates in

为艺术家提供了形式和内容，身体的多元变化形式、形体造型，尤其是通过秀场，贡献给人们大量错综复杂的形式和视觉关系。而作为内容的身体形态及形式，有助于表达文化价值观，包括宗教、政治信仰以及关于个人和社会身份的观念。身体性借此获得了新的情势，当代艺术家通过强调身体的物质性和触觉性而彻底改造了造型艺术，进入视觉文化表征，使其重新焕发生机。在对身体主题的考察中，李薇也实验运用了一系列策略与主题，例如将衣饰虚拟为身体的意象，使身体与山水、云气、光影及自然生态同化，将男性、女性身体的符号能指朦胧化、陌生化、光影化乃至行为化，并且运用发丝等身体意象，象征身体的触觉、本能的在场性质感、踪迹、印记及其感受和处境，借此转化"身体作为战场"的激进女性主义身体政治话语。而从艺术当代性话语来考量，这种实验向度依然属于身体的意象审美，意象审美尽管在媒介扩展上能够部分地逸出"美"的古典范式，但毕竟粘连于那个辖域，且会流于支配性概念。那么，身体性的秀场及秀场之当代性实验能否创造出潜在的、异类元素的、亦为处境化亦为全球化的综合体？

the fashion show arena. Through body movements, performances, spatial arrangements, a sense of presence, interactivity, surprising juxtapositions, connections, continuity, and open-ended interpretations, it conveys an ethereal and unrestrained worldly sentiment, embodying the freedom and allure that modern people yearn for. Her works and exhibition show blend, dialectically transform, and mutually shape the dynamics and stillness, the virtual and real, the distant and intimate, the visual and expressive, the classical and avant-garde. Her large-scale, continuous murals and folding screens serve as interpretive lenses for the exhibition, vividly and skillfully elucidating the charm of ink landscape painting, immersing viewers in a state of forgetting oneself and transcending reality. It is an experience that captivates and transports individuals, making them reluctant to leave during their spiritual journey.

One important point to note is that embodiment serves as both the implicit theme of this exhibition and a dimension of Li Wei's transition from academic art to experimental art. The body, or corporeality, as an artistic and ethical domain, transcends conventional art mediums. Therefore, it can be associated with the concept of "rhizome" by French philosopher Gilles Deleuze. The term "rhizome" describes interconnectedness in research and thought. It signifies the absence of a fixed starting point or endpoint, and the rejection of rigid and hierarchical organizational forms, and dominant concepts. Additionally, it possesses the capacity to link heterogeneous elements together. In this sense, the body or corporeality serves as the "rhizome" of contemporary art, as many knowledge systems of visual representation and interpretation today are fluid, non-hierarchical, non-linear, and decentralized. The body provides artists with both form and content. The diverse and ever-changing forms and body shapes, especially through clothing, contribute a plethora of intricate visual relationships with people. As for the content, the embodiment and forms of the body help express cultural values, including religious and political beliefs, as well as ideas concerning personal and societal identities. Through this, corporeality has gained new significance. Contemporary artists have thoroughly transformed the art of sculpture by emphasizing the materiality and tactility of the body, allowing it to enter the realm of visual cultural representation and revitalizing it with newfound vitality. In the exploration of the body theme, Li Wei has also experimented with a series of strategies and themes. For example, he metaphorically represents clothing as the embodiment of the body, blending the body with landscapes, clouds, light, shadow, and natural ecology. She obfuscates, unfamiliarizes, abstracts, and even enacts the signifiers of male and female bodies, utilizing body imageries such as hair strands to symbolize the tactile nature, instinctual presence, traces, imprints, sensations, and situations of the body. Through these approaches, he transforms the discourse of radical feminist body politics that conceptualizes the "body as a battlefield." From the perspective of contemporary art discourse, this experimental dimension still belongs to the aesthetic imagery of the body. Although the aesthetic imagery partially escapes the classical paradigm of "beauty" through medium expansion, it is nevertheless bound to that realm and can easily fall into dominant conceptual frameworks. So, whether the corporeal stage, as an experiment in contemporary, able to create a potential, diverse, and hybrid entity that is both situated and globalized?

苏丹
清华大学美术学院教授、博士生导师、评论家

李薇是一位行走在两个专业之间及两个领域相交边缘的实践者，她的作品有时仅以材料的形式现身，或垂范于高墙之前，或高悬于厅堂之上。但却如中国的传统艺术，将一个主观认知与感悟的世界呈现于观者面前，笔挥墨尽之处刻露出大自然中云霜冰雪之貌，风荷雨叶之声。有时这些材料也会在她的手下出神入化般收敛、张弛、回转、荡漾，更有时还会追随着舞者的身姿化作耸然而高升的烟云抑或窈然而深藏的迷雾，表现出飘逸、高渺、玄幽、悠远的意境，充满了东方式的诗性。李薇虽然是一名活跃在服装领域的设计师和教育者，但其代表作品却难以按照职业和技术归类。其作品中不仅拥有绘画的语言，也融合了塑造的手法，当身体介入其中并翩然行进时，还会产生富有节奏的律动，演化为先锋性的"行为艺术"。近年她甚至在尝试将信息技术融合在其作品中，以实践一种充分表达的可能性。长久以来，她着重在面料表现语言上进行着持续探索。丝和纱是其多年来选用的主要材料，这两种传统材料之轻柔、迷蒙的气质与艺术家想要表达的意境是统一的。同时，其作品造型的手法选择了高度的抽象性和强烈的表现性语言，形成自己独特的造型方式。以我本人对李薇的了解，其作品在创作中一定寄托了许多美好的意愿。但对于观者而言，这些如云一般翻卷，如烟一般虚幻，如闪电一般惊艳，如火一般炙热的图像，以及身体和材料之间表达的若即若离的关系，还是留下了不绝如缕的记忆与回响。

Su Dan

Professor of the Academy of Arts and Design,
Tsinghua University, Doctoral Supervisor, Critic

Li Wei is a practitioner who navigates between two disciplines and intersects at the boundaries of two fields. Her works sometimes manifest solely in the form of materials, hanging gracefully before towering walls or suspended high above halls. However, akin to traditional Chinese art, it presents a subjective realm of understanding and perception before the viewers. With each stroke of the brush and every drop of ink, it reveals the appearance of clouds, frost, snow, the sound of wind, and lotus leaves in the rain. Sometimes, these materials, under her skillful hands, converge, expand, twist, and ripple like a mesmerizing dance. At times, they transform into billowing smoke rising high or into misty veils deeply concealed, following the movements of a dancer, showcasing an ethereal, lofty, mysterious, and distant artistic conception, brimming with the poetic essence of the East. Although Li Wei is an active fashion designer and educator, her representative works defy categorization based on profession and technique. Her works not only possess the language of painting but also integrate sculptural techniques. When the body becomes involved and gracefully moves within them, they generate a rhythmic motion, evolving into avant-garde "performance art". In recent years, she has even been experimenting with integrating information technology into her works, exploring greater possibilities for expression. For a long time, she has been focusing on continuous exploration of fabric expression. Silk and organza have been her main materials of choice over the years. The delicate and hazy qualities of these two traditional materials align with the artistic mood she intends to convey. At the same time, her artistic approach in shaping the works embraces high abstraction and strong expressive language, forming its own unique style. Based on my understanding of Li Wei, her works undoubtedly harbor many beautiful intentions in their creation. For the viewers, however, these images are like rolling clouds, ethereal like smoke, thrilling like lightning, and blazing like fire. The elusive relationship expressed between the body and the material leaves behind a lingering memory and resonance.

张敢

清华大学美术学院教授、博士生导师、评论家

服装设计可以理解为以人体为载体、以织物为媒介进行的艺术创造。它融汇了时代、社会、心理、审美、科技、种族、性别、地域等多方面的因素，因此，一位优秀的服装设计师一定既能把握时尚的潮流，又有对文化的深刻理解。李薇就是这样一位设计师。当那些清秀的中国女孩穿着李薇设计的服装从你面前翩然而过时，你一定能体会到一种"中国韵致"。

全球化是个无法抗拒的潮流，人们的生活习惯和生活环境正日益趋同，与日常生活密切相关的服装也在这股洪流中变得逐渐丧失了自己的文化个性。很多敏感的设计师都意识到应该在设计中体现中国特色，但由于缺乏对中国文化的深入了解，往往流于对中国符号的堆砌和滥用。李薇是敏感的，更多了一份对中国文化深挚的热爱和理解。她尝试从传统文化入手，去挖掘可用于服装设计的灵感和元素。当一滴墨汇入清水中，慢慢地扩散晕开，留下优美而难以捉摸的线条和图案。中国画中墨色的无穷变幻也是在控制与偶然的效果中生成的。李薇从中国的水墨画中获得了灵感，而且找到了一种非常恰当的承载这种水墨效果的织物——水纱。水纱这个名字是李薇起的，因为这种用生丝手工织成的薄纱有着极好的通透性，恰似一泓清水。

李薇用水纱设计的服装宽松飘逸，当模特走动时，仿佛水墨在空间中流淌、弥散。比如她的作品《夜与昼》《韵》和《水墨戏》，都将这种韵味传达得淋漓尽致。她创作的《汉字》和《书写》系列服装作品，灵感显然来自中国的书法，连模特的发型都精心地设计成墨线的效果。

李薇的作品呈现了中国艺术特有的含蓄和飘逸，但同样不乏现代感。在造型上，她设计的旗袍和裙装，有的非常简洁，有的则充满了质感的冲突和对比，如《清·远·静》和《基因》。在视觉效果上，李薇用通透的水纱凸显了身体曲线的曼妙，这显然与中国的传统观念相去甚远。因此，李薇的服装虽然有明显的中国韵味，但它们的语言是国际化的、开放的。李薇的创作领域很广，并不囿于服装设计，她的染织作品《月光》《清远静》《行云流水》完全可以被看作独立的艺术创作，就像一幅幅抽象的水墨画，空灵淡远。

观看李薇的作品，观众们一定能感受到一股清新的中国气息，但绝不会有任何隔阂之感。人们对美的追求是相通的，李薇的作品恰好为观众提供了一个欣赏和理解中国艺术的机会。

Zhang Gan
Professor of the Academy of Arts and Design,
Tsinghua University, Doctoral Supervisor, Critic

Fashion design can be understood as an artistic creation that uses the human body as a carrier and fabric as a medium. It integrates various factors such as era, society, psychology, aesthetics, technology, race, gender, and region. Therefore, an outstanding fashion designer must be able to grasp the fashion trends and have a deep understanding of culture. Li Wei is such a designer. When those delicate Chinese girls pass by you wearing Li Wei's designed clothing, you can definitely experience a kind of "Chinese charm".

Globalization is an irresistible trend, as people's habits and living environments are increasingly converging. Clothing, which is closely related to daily life, is gradually losing its cultural individuality in this powerful current. Many sensitive designers have realized the importance of embodying Chinese characteristics in their designs. However, due to a lack of in-depth understanding of Chinese culture, they often resort to piling up and abusing Chinese symbols. Yet Li Wei is sensitive, with a deeper love and understanding of Chinese culture. She attempts to start from traditional culture and explore inspirations and elements that can be used in clothing design. Just like a drop of ink merging into clear water, it slowly spreads and diffuses, leaving behind beautiful and elusive lines and patterns. The infinite variations of ink in Chinese paintings are also generated through a combination of control and accidental effects. Li Wei drew inspiration from Chinese ink paintings and found a suitable fabric to embody the ink effect - aqua fabric (shui sha). Li Wei named this fabric "aqua fabric (shui sha)" because it is made of raw silk through manual weaving, and it has excellent transparency, resembling clear water.

The clothing designed by Li Wei using aqua fabric is loose and flowing. When the models walk, it appears as if the ink is flowing and dispersing in space. For example, her works "Night and Day," "Melody," and "Ink Play" perfectly convey this charm. Her artwork series "Chinese Characters" and "Calligraphy" are clearly inspired by Chinese calligraphy, even the models' hairstyles are carefully designed to resemble ink lines.

Li Wei's works embody the inherent subtlety and elegance of Chinese art, while also exuding a sense of modernity. In terms of styling, some of her cheongsams and dresses are very simple, while others are filled with contrasting textures and conflicts, such as "Qing Yuan Jing" and "Genes". In terms of visual effects, Li Wei highlights the exquisite curves of the body with transparent water gauze, which is clearly different from traditional Chinese concepts. Therefore, although Li Wei's clothing exhibits distinct Chinese charm, their language is international and open-minded. Li Wei's creative scope is broad and not confined to fashion design. Her dyed and woven works such as "Moonlight," "Qing Yuan Jing", and "Flowing Clouds and Running Water" can be regarded as independent artistic creations, resembling abstract ink paintings, ethereal and distant.

Li Wei's works are being exhibited this time, and I believe that the audience will be able to feel a fresh Chinese charm without any sense of distance. The pursuit of beauty is universal, and Li Wei's works precisely provide an opportunity for viewers to appreciate and understand Chinese art.

许平
艺术家、中央美术学院研究生院院长

李薇的作品散发着飘逸、向外的感觉,追求超过服装面料之外本身的东西,引人深思。作品可以忘形,和艺术家不一样,可以在片刻之间也许真的就会寻找到一些东西。她能把第一感觉始终控制在过程里面,不要成为工匠的那种,要得意忘形。另外,她的作品隐约带有一点毕加索的那种味道。

《基因》这一设计作品更难表现,有形可循反而会容易设计。李薇的长处还是基于时装这一领域,一旦走得太远可以回头看看,是不是所擅长的那一部分。不管是做服装还是做时装艺术,其底座还是人的身体的感受,以及身体带来的联想、人文环境,我觉得这是每个人自己特别重要的因素,不要轻易忽略。从美学的角度来讲,崇尚力量的美学和崇尚优雅的美学其实是两种不同的力量,有时,我们会过于偏重崇尚力量的美学,使其过于封闭,而忽视了优雅的部分。其实以柔制胜,倒是未必克刚,以柔的表现,表达其独特的空间。

Xu Ping

Artist, Dean of the Graduate School of the Central Academy of Fine Arts

Li Wei's works have a sense of elegance and outwardness, pursuing something beyond the clothing fabrics themselves, which is thought-provoking. Her works can be ecstatic, unlike ordinary artists, she might really find something in a moment. She is able to control the first impression throughout the process, not becoming like a craftsman, but being lost in ecstasy. Additionally, her works have a taste similar to Picasso.

The "Genes" is a more difficult piece to portray, as having something tangible to grasp can actually make it easier to design. Li Wei's strength lies in the field of fashion, and once she has ventured too far, she can always look back and see if it aligns with her expertise. Whether it is fashion or fashion art, the foundation is still based on the perception of the human body, as well as the associations and cultural environment that the body brings. I believe this is a particularly important factor that should not be easily abandoned. From an aesthetic perspective, the pursuit of aesthetics of power and elegant aesthetics are actually two different forces. There was a time when we leaned towards the aesthetics of power a bit too much, closing ourselves off and weakening the elegance. In reality, victory can be achieved through gentleness rather than brute strength. The expression of gentleness is equally important and it has its own space.

高岭
中国著名艺术批评家

衣者，天地形——走进李薇的艺术世界

李薇的衣饰不仅是为人的，更是为穿着者与天地产生连结；李薇的服装设计面料是为衣饰的，更是向自然致敬的。要走进著名服装艺术家李薇女士的艺术世界，理解她的衣饰和面料背后的文化意义和精神实质，天、地、自然是最重要的参照。

李薇擅长用绡和水纱，在她眼中，它们能够传递出大自然的无限生机，能够体悟到周遭的温度——她用模特着衣的方式来塑造大自然拟人附体的衣者风范，她用材料铺展的方式来捕捉天地间云蒸霞蔚的意象气质。我们在她二十年前开始成形并不断拓展的《夜与昼》《青绿山水》《韵》《壁影》和《荷塘月色》等系列作品中，看到的不是对身体的生理属性的彰显，而是对身体的文化属性的澄明。纱薄，不意味裸露；衣透，不暗示情色。与自然谐，同造化生，源自自然又复归于自然，才是这一批批服饰作品的诉求所在。

李薇的衣饰平展开来，形制多似鸟翼，它们或披挂于身，或悬挂于厅堂之上，与其说是为了暴露曼妙的身材，毋宁说是为了带动衣者和观者的身心去亲身体悟大自然的风情与气韵——它们扶摇而上，飘逸轻盈，自在又自由，出自远古，附身天地，实乃天作人和。

李薇对于服装艺术的贡献，在于她深刻理解并且把握当代时装潮流以人为本的同时，又将这股潮流引入对本土文化的深刻认知上。服装包括对服装的艺术表达，当然离不开人以及人的文化，但人以及人的文化，又离不开其赖以生存并取之不尽的自然造化，没有人的文化是难以理解的，而脱离了自然的人及其文化更是不可能的。让为人的服装文化传递出自然的气息，折射出造化的身影，甚至反映出天地的精神，是当代文化重拾人的身心合一、知行一体后的一次提升，是主观的人的文化与客观的自然的实在的一次全方位的契合。她的《清·远·静》的大型壁挂作品，她的《空山鸟语》和《韵致》作品中的材料实验，她的《山峦》和《风声》纤维组画和屏风，即是她二十年服饰设计和材料研发方面坚持不懈努力的最终呈现。作品都清晰并且有力地说明，真正的服装设计师，最终都将是服装文化的哲学思考者，终究都将于有形有色的空间实体中，见出无形无色的生命之本源。

如是，李薇的薄衣便具有了生命存在的意义，因为自然固然亘古永恒，对应着时间的无限性，不为人事所累，

Gao Ling
Famous Chinese Art Critic

Clothing: the Reflection of the Origin, Entering Li Wei's Artistic World

Li Wei's fashion design is not only for people but also serves as a path connecting humanity to the realm of heaven and earth. Li Wei's fabrics are not only for clothing but also pay tribute to nature. To truly understand the cultural significance and spiritual essence behind the clothing and fabrics of renowned fashion artist Li Wei, it is crucial to reference the natural world of heaven and earth.

Li Wei excels in using raw silk and aqua fabric. In her eyes, they can convey the boundless vitality of nature and allow her to perceive the ambient temperature. She shapes the clothing on the models to embody the essence of nature personified. Through the draping of materials, she captures the ethereal and magnificent imagery between heaven and earth. In her artwork series such as "Night and Day" "Jade Green Landscape" "Rhythm" "Silhouette" and "Moonlight over the Lotus Pond", which have been shaping and expanding for the past twenty years, what we see is not the explicit display of the physical attributes and carnal desires of the body, but rather the clarity of its cultural significance. Thin fabric does not imply nudity; translucent clothing does not suggest sensuality. Harmonizing with nature, being part of the creation, originating from nature and returning to nature, is the true essence and appeal of these batches of clothing creations.

Li Wei's fashion designs are often reminiscent of bird wings when spread open. They can either be draped on the body or suspended in exhibition halls. Rather than being intended to expose a perfect figure, they are meant to inspire both the wearer and the observer to personally experience the charm and elegance of nature. They soar gracefully, light and ethereal, embodying a sense of ease and freedom, the fusion of heaven and earth, and a true harmonization between humanity and nature.

Li Wei's contribution to fashion art lies in her profound understanding and grasp of contemporary fashion trends that prioritize the individual, while also incorporating this trend into a deep appreciation of local culture. Clothing, including the artistic expression of clothing, can not be separated from people and their culture. However, people and their culture are inherently intertwined with and reliant on the inexhaustible creations of nature. It is difficult to comprehend culture without considering the presence of humanity, and it is impossible to detach human culture from nature. Allowing the clothing culture of humanity to convey a natural breath, reflect the presence of creation, and even embody the spirit of the universe is a contemporary cultural elevation that restores the unity

不被社会历史所限，但它的不朽是因为有人的存在。在人之外的世界，即使是无限而且无穷的，也是无意义的，因为无人的世界便无言说，无言说则无意义。

如是，李薇的轻纱在人的世界里，以有限但不懈勤勉的智慧，努力接近自然的超越性，如秉承天人合一的原则，以人应天，以跟随时代的艺术形式，接应、映射自然之不朽，凸显人的艺术的超现实性同样可以不朽。

衣者，人也，为自然造，为天地形。

of body and mind, and the integration of knowledge and action. It signifies a comprehensive harmonization between subjective human culture and objective nature. Her large-scale mural, "Qing Yuan Jing", her material experiments in "Birdsong in the Empty Mountains" and "Elegance", and her fiber art group and screens of "Mountain Ranges" and "Whisper of the Wind" are the ultimate culmination of her unwavering efforts in fashion design and material research over the past twenty years. They clearly and powerfully demonstrate that a true fashion designer ultimately becomes a philosopher of fashion culture, revealing the intangible essence of life within tangible and colorful spatial entities.

Indeed, Li Wei's delicate clothing takes on the meaning of existence because while nature itself is enduring and eternal, transcending the limitations of time and unaffected by human affairs or societal history, its immortality is attributed to the presence of humanity. In a world devoid of human presence, even if it is infinite and boundless, it becomes meaningless, for a world without human expression lacks significance and purpose.

Thus, Li Wei's sheer fabrics in the human world strive, with limited yet unwavering wisdom and diligence, to approach the transcendence of nature. Guided by the principle of unity between heaven and humanity, they adapt to the ever-changing artistic forms of each era, responding to and reflecting the immortality of nature, highlighting the surrealistic nature of human artistry, which can also be everlasting.

The clothing represents humanity, created by nature and shaped by the universe.

姜绥祥

香港理工大学教授、博士生导师、艺术家、设计师

再传统，再中国 —— 李薇的艺术设计

厚重绵长的文化积淀与纵情洒脱的艺术创作之间碰撞出的火花，是李薇从"去传统"到"再传统"，从"去中国化"到"再中国化"的无尽思考。与同代服装设计师一样，李薇独特的个人经历赋予了她的作品强烈的寻找文化之根和表达自我的情怀。从李薇的作品中不难发现，从20世纪90年代开始她的创作就带有探索东方艺术本质的倾向，走上了一条寻找新载体来承载中国文化精神的道路。

李薇是用服装设计师的眼光去创作纤维艺术的践行者。她创造性地将纤维材料与绘画及工艺手段一起综合考量开展艺术实践。《清·远·静》是李薇在纤维装置创作方面的代表性作品。结合织绣和染画的再创造，纤维材料兼具水墨绘画的语言和意境。裁剪冰绡，轻叠数重，潇洒恣意的青烟墨雨追寻一种水墨交融的梦幻般的诗性空间。凭借对人与天、地自相协调东方美学的感悟，作品得以超越二维空间，三位合一达成意蕴深远的虚拟世界。凭一个服装设计师对维度的把握，乃至对视觉模糊性的深刻认知，李薇的作品才能达至如此出神入化的弥漫幻境。

李薇是以纤维艺术的观念来创新服装设计的先行者。她的服装设计从传统服饰文化入手，进而在纤维材料与服装创作的承接与转换关系上已然超越了时空维度束缚。从纤维艺术置换到服装设计，作品《夜与昼》及《青绿山水》不但凸显出李薇创作服装驾轻就熟的深厚功力，同时也呈现了她无尽的艺术才情，可谓骨感不无轻盈，深沉近乎虚空，冷峻兼有妖娆。不管是纱还是绡，经过材料再造，她都能在服装这种载体上剪裁出东方元素和中国文化的意境。李薇先运用新材料、新手段在服装面料上进行纤维艺术创作，再而服装，无疑是挑选了一条艰辛也极具挑战性的设计道路。

无论是纤维艺术还是服装设计，都是李薇对中国文化与现代美学、当代设计语言融合的探讨，也是对自己想要表达的"天人合一"的东方美学精神的致敬。作品如其人，豪放而不失文气，李薇追求高远的品格亦赋予她的艺术以恒久的魅力。

Kinor Jiang
Professor, PhD Supervisor, Artist and Designer,
Hong Kong Polytechnic University

More of Tradition, More of China: The Art Design of Li Wei

The sparks, as the result of collision between a profound and time-honored culture with free and unconstrained art creation, are really the contemplation of Li Wei in a process of dissimilation from de-tradition to re-tradition and from de-China to re-China. Like other contemporary fashion designers, the unique personal experiences of Li Wei render her works a strong feeling of root-finding and self-expression. From her works it is not hard to discern that since the 1990s she has started to search the essence of oriental art and to explore a new way to carry the Chinese cultural spirit.

Li Wei is a practitioner who creates fiber art from the view of a fashion designer. She undertakes her art practice by creatively and comprehensively taking into consideration fiber material, painting, and technique. "Qing Yuan Jing" is a representative work of Li Wei in fiber art. Integrated with the re-creation of embroidery and dyeing, the fiber materials are rendered the elements and style of an ink painting. Magically cut, and gently placed, the created scenery with ink saturates a dreamy and poetic space. With her thorough understanding of oriental aesthetics founded on the harmony of human, heaven, and earth, her works break the two-dimensional space, and form an imaginary world rich in meaning. Armed with mastery of dimensions of a fashion designer, with the profound cognition of visual ambiguity, Li Wei makes her works a realm of fantasy.

Li Wei is a pioneer in designing apparel from the perspective of fiber art. Her fashion design starts from the traditional culture of costume, but surmounts the dimensions of time and space in the transition from fiber materials to apparel creation. Fiber art is integrated into fashion design. The works "Night and Day" and "Turquois Landscape" exhibit her sophisticated skill in fashion design and her endless artistic passion. Pointed yet fluent, profound as if empty, calm and cool yet enchanting. Yarn or silk, by restructuring, she can always tailor a garment of oriental elements and of Chinese culture. With new materials, new techniques, Li Wei starts with creation of fiber art on the fabrics, then advances to fashion design, which is a hard and challenging road.

In both fiber art and fashion design, Li Wei is exploring the integration of Chinese culture, modern aesthetics, and contemporary design. Her works exhibit her desire to express the oriental aesthetic spirit of harmony between human and nature. Works reflect personality. Free yet gentle, the pursuit of Lei Wei renders her art a long-lasting charm.

国外学术评价

克莱尔·帕雅茨科夫斯卡
前英国皇家艺术学院材料与时尚专业教授、电影制作人

了解其他的物质性

皇家艺术学院 Entrance Gallery 的展览邀请到北京清华大学美术学院纺织系教授、艺术家、设计师李薇。本次展出的作品是从其过去二十年的作品当中精选而出,包括编织、染色、绘画、裁剪、缝纫等服饰作品。

时尚的悖论是时间和永恒。我们所熟知的时尚文化是对服装廓型、裁剪或垂褶的创新运用。服装和面料间的关系是维续时尚复杂循环能量流通的途径之一。作为设计师、工匠和制作者,李薇在寻找服装面料间的微妙关系,以及设计师和材料之间不可分割的紧密联系。材料文化研究学院有幸邀请李薇参加本次展览,由她主持丝绸研讨会并就其创作实践开设讲座。

作为欧洲的一部分,英国有着悠久的东西方文化交流历史。早在两千年前,古罗马人把中国人带到伦敦生活,丝绸贸易给腓尼基人和古希腊人带来了丰富的知识和材料文化。兰州是黄河沿岸的城市和古丝绸之路的贸易点,在兰州博物馆里一尊雕像展示了希腊神话中神奇的"金羊毛",实际上是一种由真丝织成的奇妙的发光材料。丝绸光泽亮丽、质地细腻,不需要媒染剂和单宁就能吸收染料和颜料,是沿贸易路线买卖最有价值的商品之一,在西方,丝绸面料具有货币价值。随着纺织和编织技术的发展,与其相关的其他种类的知识,如造纸、数学、哲学、工程学、食品、香料以及另一种美学也随之产生、衍变。

十八世纪末,欧洲在英国工业革命之前引领了科学思想的启蒙运动,但这是以纺织品贸易路上知识的自由流动为代价的。人类学家克劳德·列维·施特劳斯(Claude Levi Strauss)在他的论文《神话与意义》(Myth and Meaning)中指出,欧洲热衷于颂扬科学方法中新发现的理性主义,以至于其哲学变得模式化,在新力量的概念与"他者"之间强加了二元对立。古老的智慧迅速被斥之为"迷信",被诋毁为"史前"或原始。大约两个世纪后,我们可以摆脱这一传统的束缚,重新思考纺织品思维的"软逻辑"范式。

十八世纪的数学家、哲学家莱布尼兹是一个创造性天才中的例外,他来中国旅行过,是中国文化的崇拜者。他的数学是以法国哲学家吉尔·德勒兹(Gilles Deleuze)的逻辑进为基础,并进行后现代的颂扬。褶皱是一种结构,它强调反身性,而反身性又是一种意识,是一种曲线运动,它与僵硬的网格几何图形和坚硬的物质性是对立的。德勒兹在书中写到了褶皱的复杂性,认为纺织品思维是对传统工业及工程的一种丰富。重新思考传统权利、挑战新的确定性的哲学很少受到欢迎。自十九世纪中叶以来,"另类"的软逻辑一直被视为一种"弱点",是一种女性特质或矩阵偏差,与安全的防御性几何结构格格不入。通过创造性艺术和设计实践的物质性特权,我们学习莱布尼兹数学、包豪斯设计师列维·施特劳斯的结构主义等大胆的欧洲倡议,

International Academic Comments

Claire Pajaczkowska
Former professor of Materials and Fashion at the Royal College of Art,
film maker

Knowing Other Materialities

The exhibition at RCA Entrance Gallery is by artist and designer Li Wei, a professor of the Academy of Arts and Design in the Department of Textiles and Fashion design at Tsinghua University Beijing. It is an exhibition of works in weave techniques, in dyeing and painting, in pattern cutting and tailoring and in fashion design. These works are selected from a considerable archive and current collection of pieces produced over the past two decades.

A paradox of fashion is its simultaneous timelines and timelessness. Innovations in silhouette, cut or drape generate the play of call and response that we know as fashion culture. The relationship between garment and fabric is one of the main arteries that maintains the circulation of energies in the complex cycle of fashion. Li Wei has explored this relationship as a maker and artisan ad well as a designer, and these works celebrate the pleasure of that close, intimate relationship that designers have with materials. It is a privilege for the School of Material research culture to invite Li Wei to exhibit her works, to lead workshops on draping in silk, and to offer illustrated talks on her creative practice.

As a part of Europe the UK is part of a long history of the exciting cultural exchange between Occident and Orient. Two thousand years before Romans brought Chinese people with them to live in London the silk trade had brought a rich culture of knowledge and materials to the Phoenicians and Ancient Greece. In the archaeological museum of Lanzhou a city on the Yellow River and trading point in the old silk road a bornze figurine shows that the Greek myth of the magical "golden fleece", was, in fact that wonderful luminous material that is woven silk. Lustrous is sheen, miraculous in delicacy, and extraordinary in its ability to take dyes and pigments without need for mordants and tannins, silk was one of the most valuable commodities bought and sold along the trade rout, and so unchanging was its value to the West that bolts of woven silk textile were traditionally used as currency itself. Along with textile and weaving skills arrived other kinds of knowledge, paper making, mathematics, philosophy, engineering, foods, spices and encounters with another ethic aesthetic.

When Europe led the Enlightenment of photo scientific thinking that preceded the UK industrial revolution on the late eighteenth century, it did so at a cost to the free flow of knowledge exchanged along the textile trade route. Anthropologist Claude Levi Strauss, in his essays "Myth and Meaning" points out that Europe was so keen to celebrate the new found rationalism of scientific method that its philosophies became schematic in imposing a binary opposition between the concepts of the new power and their "others". Old wisdoms were rapidly dismissed as ;superstition, denigrated as "prehistoric" or primitive. Some two centuries later we can release ourselves from the strictures of this legacy

以及其他东西方之间可以产生真正灵感碰撞的知识。

李薇对编织和数字纺织品形式的实验在许多方面都具有独创性。对皇家艺术学院2016年的研究而言，其作品中将染色与悬垂相结合的部分最具意义。丝绸聚合物和其他一些纤维材料一样具有特殊的"吸湿"和导电性能，这不仅仅是材料的价值，也是一种观念的价值。我们热爱色彩的亮度和丰富性，这一直是我们的乐趣所在。然而，我们现在更感兴趣的是这种包容度和容纳能力如何为概念化权力提供新的范式。研究科学、技术和工程学（STEM）议程的限制在于人们不加思索地遵从"影响"理念是可衡量的影响他人行为的力量，似乎知识的作用只能通过它改变他人行为的程度来衡量。接收、吸收、流动和关系的能力是另一种矩阵，是"影响"的基质，而纺织品的物质性则是这种基质的具体先驱。李薇的作品正是以这种能力展示了其慷慨和美丽。后数字经济时代的文化随着新的信息高速公路的到来，成为横跨东西方的连接网络，就像新的纬度曲线和放射线。新探险家、探寻未知领域和海洋深处的艺术家和设计师们，为旧有的、枯竭的确定性带来了新的知识。李薇教授的纺织艺术与 RCA 材料学院的研究文化的相遇，为我们提供了一个探索不同知识临界点所产生的能量的机会。

英国皇家艺术学院（RCA）在 2016 年再次成为世界领先的艺术与设计大学，不仅仅是因为它接受跨文化和跨学科的交流，而这种交流正是真正原创性的创意产生背景。与未知的相遇是点燃隐性知识与显性知识之间联系的火花。与材料的相遇，是不同地域和空间之间进行文化和符号探索的对话。

李薇的作品启发我们重新思考我们在悬垂、固定、连接、流动、吸湿、染色中所发现的创意。

and rethink the paradigms of the "soft logics" of textile thinking. Eighteenth century mathematician philosopher Leibniz, the exception that is the rule in creative genius, was a Sinophile traveller to China and great admirer of the culture. His mathematics is the basis for French philosopher Gilles Deleuze' postmodern celebration of the logic of "the fold". The fold is a structure that privileges reflexivity which is held consciousness, the curvilinear movement that is the antithesis of rigid grid form geometry of hard rigid materiality. Delouse writes of the complexity of the fold and suggests that textile thinking is a radical enrichment of traditional industrial engineering. The philosophy that rethinks traditional entitlenebt and challenges new certainties is rarely popular. The soft logic of "otherness" has, since the mid nineteenth century been considered as a "weakness", a femininity or matrixial aberration from the safety of defensive geometries. With the privileged materialities of creative art and design practice we are invited to learn from the bolder European initiatives of Leibniz mathematics, the Bauhaus designers Levi Strauss structuralism and others' real inspired encounters between East and West.

Li Wei's experiments with woven and digital textile forms claim originality in many ways . It is the aspect of her work which combines dyeing and draping that is of the most significance to RCA research in 2016. The quality of silk polymer, and some other fires, to have exceptional properties of "wicking" and conductivity are of conceptual as well as material value. We love the richness and intensity of color and luminosity that has always been a source of pleasure to us. However we are also now more interested in the way that this capacity of absorbency and; containment' offers new paradigms for conceptualizing power. The strictures of a STEM agenda for research is the damage caused by unthinking compliance with its ideas of "impact" as measurable power to affect the activity of others, as if the agency of a knowledge can only be measured by the extent to which it alters the agency of others. The capacity for reception, absorption, streaming and relational ;its is another matrix, the substrate of impact, which is specifically pioneers by textiles materiality. It is this capacity that Li Wei's works display with generosity and beauty. The post digital economies of culture that arrive with the new information highways are networks of connectivity that cross East and West like new curvilinear and radial lines of latitude. New explorers of new empires, the artists and designers that sound the uncharted territories and oceanic depths bring new knowledge to old legacies of exhausted certainties. This encounter between Professor Li Wei's textile art and the research culture of the RCA School of Material offers us an opportunity to explore the energies generated at the threshold between different knowledges.

The RCA is, again in 2016, the world's leading university of art and design not least because it welcomes the intercultural and interdisciplinary exchanges that are the context for the creative ferment of real originality. Encounters with the unknown are the spark that ignites that connection between tacit and explicit knowledge. The encounter with materials sets into play the dialogue between the different territories and spaces of cultural and symbolic adventure.

Li Wei's works invite us to rethink the creativities we find in draping, holding, connecting, streaming, wicking, dyeing.

媒体评价

意大利和比耶拉的媒体

在意大利比耶拉（Biella）城市艺术中心举行的《视·听——李薇艺术作品展》的报道中评价："李薇的作品立足于国际，把东方传统美与艺术、纤维相结合（News Biella），她的服装作品十分轻盈，微妙的颜色几乎让人感触不到，相当精致（La Stampa）。

克莱尔·帕雅茨科夫斯卡
著名策展人、英国皇家艺术学院教授

在2016年皇家艺术学院"李薇个人作品展"上说："李薇教授是把时装与当代艺术融合并带到英国的第一人。服装和面料之间的关系是维续时尚发展的要道之一，作为服装设计师和纤维艺术家的李薇用她的作品完美的诠释了材料与设计及艺术之间的微妙关系。李薇教授的艺术作品与皇家艺术学院对于材料的研究为两国跨文化的学术交流提供了一个难得的机会。"

项晓炜
驻英使馆文化参赞

在皇家艺术学院"李薇个人作品展"开幕致辞中说："李薇教授的作品从中国传统水墨画中借鉴传统却又超越了传统，她向世界展示了中国艺术家的自信。"

Media Comments

The Italian and Biella media

The media in Biella, Italy, commented in the report: "Li Wei's works are based on the international, combining Oriental traditional beauty with art and fiber (*News Biella*). Her works are very light, and the subtle colors are almost impossible to touch, rather delicate (*La Stampa*).

Claire Pajaczkowska
renowned curator and Professor at the Royal College of Art

Professor Li Wei was the first person to bring fashion and contemporary art to the UK. The relationship between clothing and fabric is one of the keys to the sustainable development of fashion. As a fashion designer and fiber artist, Li Wei perfectly interprets the subtle relationship between material, design and art with her works. Professor Li Wei's works of art and the royal college of art's research on materials provide a rare opportunity for cross-cultural academic exchanges between the two countries.

Mr. Xiang Xiaowei
Minister Counsellor for Cultural Section at the Chinese Embassy to the UK

He stated in his opening speech at the "Li Wei Solo Exhibition" held at the Royal Academy of Arts: "Professor Li Wei's works draw inspiration from traditional Chinese ink paintings, yet they also transcend tradition. She showcases the confidence of Chinese artists to the world."

李薇简介
Li Wei Introduction

清华大学美术学院染织服装系长聘教授、博士生导师
当代艺术家、服装设计师
巴黎高等装饰艺术学院访问学者
中华服饰文化研究会副会长
中国流行色拼布委员会主任委员
中国电影美术学会服装造型专业委员会副主任
中国纺织工程学会时装艺术专业委员会副主任
中国美术家协会会员
中国纺织非遗推广大使

Li Wei is a tenured professor and doctoral supervisor at the department of Textile & Fashion Design of the Academy of Arts & Design, Tsinghua university. She is a contemporary artist and fashion designer. She has also served as a visiting scholar at the École des Arts Décoratifs Paris. Li Wei holds the position of deputy director at the Association of Chinese National Dress, chairman of the China Fashion Color Association, Deputy Director of the Costume Styling Professional Committee of the China Film Art Direction Academy, deputy director of the Fashion Art Professional Committee of the China Textile Engineering Society, and is a member of the China Artists Association. She is also an ambassador for the promotion of Chinese textile intangible cultural heritage.

031

李薇教授集学者、艺术家、服装设计师等多重身份于一身。从教四十年来，她一直根植于传统文化，以传统手工艺为媒介，专注于国际时尚设计与全球当代艺术的创新实践研究。通过独特的东方诗性视角，创作出极具新"东方视觉"气韵的时装艺术和纤维艺术作品。出版《中国工艺美术全集·国卷·技艺卷·刺绣 抽纱 编结 剧装道具篇》《京剧服饰制作技艺研究》《绣娘》《染匠》《织匠》《中国传统服饰图鉴》等多部学术著作；在国内外核心期刊发表论文30余篇；主持国家社科基金项目、清华大学领军人才专项及国家文化创新工程项目等多项重要科研项目；主持策划国际学术研讨会10余次；国内外重要学术研讨会做主旨发言30余次。

其作品曾多次荣获国家级奖项，2004年服装作品《夜与昼》获得"第十届全国美术作品展"金奖；2009年作品《清·远·静》获得"从洛桑到北京第六届国际纤维艺术展"金奖；2014年获得"亚太经合组织（APEC）会议领导人服装设计"荣誉奖，同年获得"中国十佳服装设计师"称号；2022年获得"北京冬奥会和冬残奥会制服设计"铜奖；2022年获得"中国纺织非遗推广大使"荣誉称号。近十年来，多次在英国、法国、美国、意大利、日本、德国、西班牙和一些非洲国家举办个人作品展，并发布植物染、高级定制、非遗主题个人服装秀10余次。

Professor Li Wei embodies multiple roles as a scholar, artist, and fashion designer. Over her forty-year teaching career, she has remained rooted in traditional culture, using traditional handicrafts as a medium to focus on innovative research in international fashion design and global contemporary art. Through a unique Oriental poetic perspective, she creates fashion and fiber art that exudes a distinct "New Oriental Visual" charm. She has published several significant academic works, including *Complete Works of Chinese Arts and Crafts - National Volume - Techniques of Embroidery, Drawnwork, Knitting, Opera Costumes and Props, Research on Beijing Opera Costumes Production Techniques, Embroiderer, Dyer, Weaver* and *Illustrated Handbook of Chinese Traditional Clothing*, among others. She has authored over 30 papers in domestic and international core journals, led numerous important research projects including National Social Science Fund projects, Tsinghua University's Leading Talent program, and National Cultural Innovation Engineering projects. She has also organized more than 10 international academic seminars and delivered keynote speeches at over 30 major domestic and international academic conferences.

She has repeatedly received national-level awards for her work. In 2004, her fashion work "Night and Day" received the Gold Award at the 10th National Exhibition of Arts. In 2009, her work "Qing Yuan Jing" earned the Gold Award at the "From Lausanne to Beijing", the 6th International Fiber Art Biennale Exhibition. In 2014, she won the Honorary Award in the "Garment Design for APEC Economic Leaders "; in the same year, she was granted the title of "China's Top Ten Fashion Designers" In 2022, she achieved the Bronze Award for "Beijing 2022 Winter Olympics and Winter Paralympics Uniform Design," and was also bestowed with the honorary title of "Ambassador of the Chinese Textile Intangible Cultural Heritage". Over the past decade, she has organized numerous solo exhibitions of her works in countries such as the United Kingdom, France, the United States, Italy, Japan, Germany, Spain, and Africa. She has also conducted over 10 personal fashion shows focusing on themes such as plant dyeing, haute couture, and intangible cultural heritage in clothing.

作为艺术表达的手段无论是语言还是图像或是声音，其本身都是有限的。但中国艺术的追求，绝不是仅仅停留在有形的声、色上，声、色只是写意的手段，而"意"的境界则是超越了声和色的。大音息声，大象无行。艺术便存在于这无形的大道之中。

——李薇

There is an end to the words, but not to the message Languages, images and sounds as ways of expression are all limited, but the goal of Chinese arts will not stay within these methods. Genuine music is hardly audible and true images is nearly invisible. Art actually lies in this intangible spirit, or in other words, the Great Dao.

———Li Wei

夜与昼

2004
绡、水纱
150cm×160cm

作品吸收了中国水墨韵味和意境，蕴有老子的"有无相生、昼夜交替、黑白相倚、阴阳流变"的哲学意味。追求单纯中的丰富、虚空中的气韵。以中国大文化为背景，以现代服饰的表现语汇，抒发了洒脱、高贵和融于天地之间的情怀。

Night and Day

2004
Gauze
150cm×160cm

The works have absorbed the charm and artistic conception of Chinese ink painting, with the philosophical meaning of Laozi's "being and not being, the alternation of day and night, the reliance of black and white, the change of Yin and Yang". The pursuit of simplicity in the rich, empty spirit. With the background of Chinese culture and the expression of modern costume, it expresses the free and easy, noble and harmonious feelings.

摄影：曹迁
Photography: Cao Qian

040

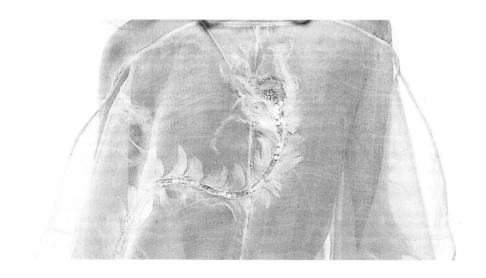

044

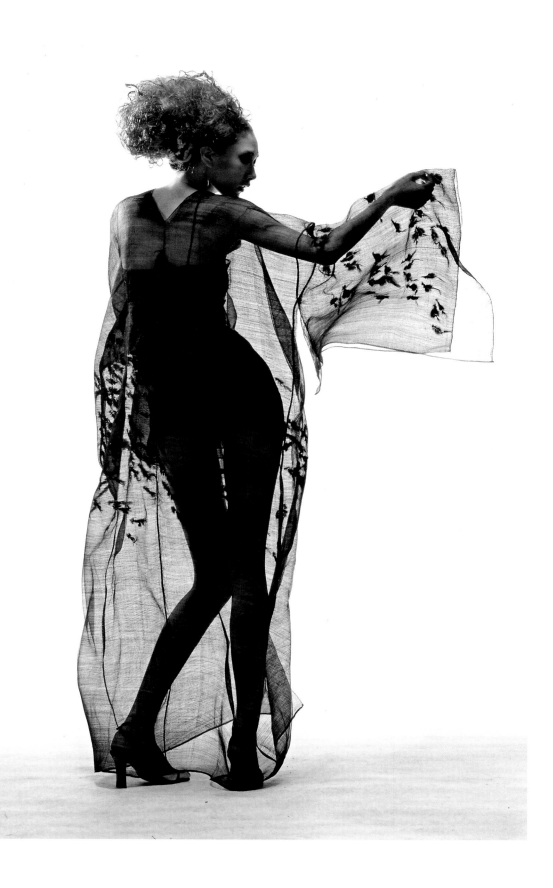

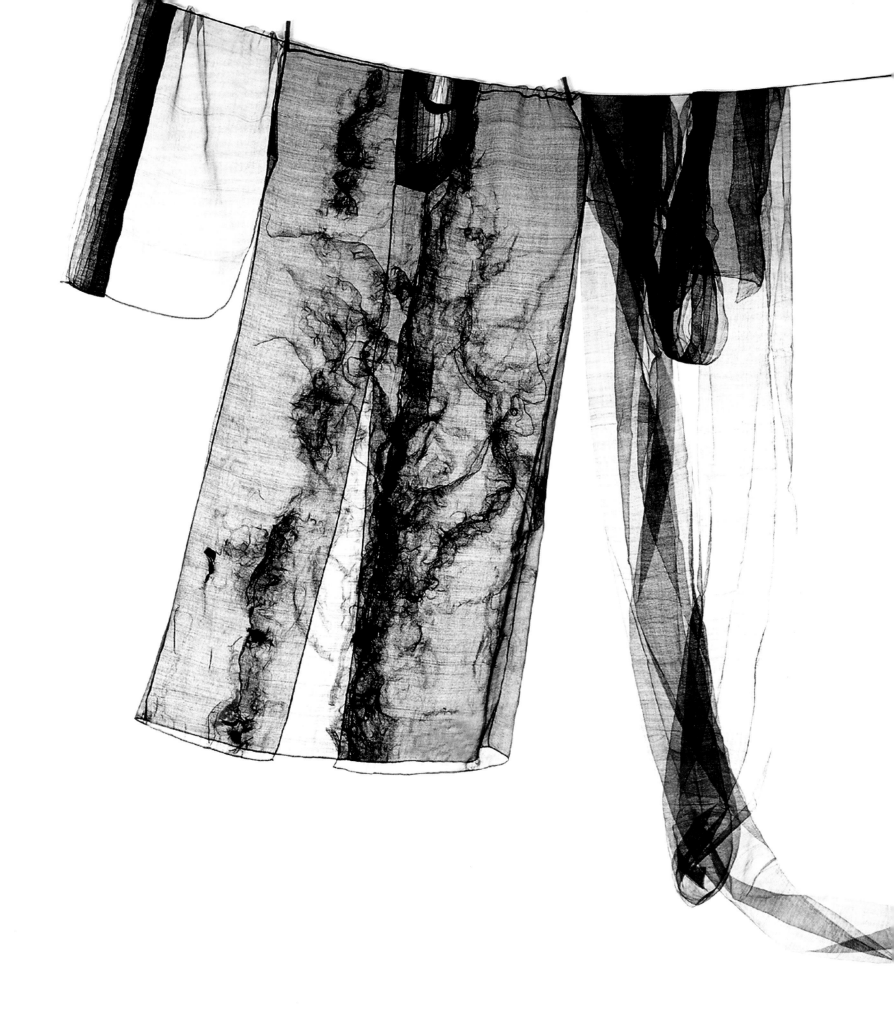

七彩云南

2021
真丝

作品以云南腾冲的地貌特征及非遗植物染等本土文化为灵感表现创意构思，凸显天、地、人和谐共生的设计理念。设计中，所用真丝面料均采用手工植物染工艺；色彩上，以赤、黄、蓝、黑、白等色为主，表现七彩云南之美；造型上，既有中国传统服装形制也有现代礼服形制，将地域文化与国际时尚相结合，呈现保山腾冲等地的自然环境与人文风貌，展现一幅从人文到自然、天地人共生的和谐画卷。

Colorful Yunnan

2021
Silk

The work draws inspiration from the geographical features and traditional plant dyeing techniques of Tengchong, Yunnan, showcasing creative concepts that highlight the harmony between heaven, earth, and humanity. The designs utilize handcrafted plant-dyed silk fabrics, employing colors such as red, yellow, blue, black, and white to portray the beauty of colorful Yunnan; In terms of style, the series incorporates both traditional Chinese clothing silhouettes and modern formal wear, blending regional culture with international fashion. This fusion captures the natural environment and cultural essence of places like Tengchong in Baoshan, presenting a harmonious tapestry that spans from human culture to the natural world, symbolizing the coexistence of heaven, earth, and humanity.

摄影：冯海
Photography: Feng Hai
化妆：姜月辉
Make-up: Jiang Yuehui
模特：张珈琦、王梓
Models: Zhang Jiaqi, Wang Zi

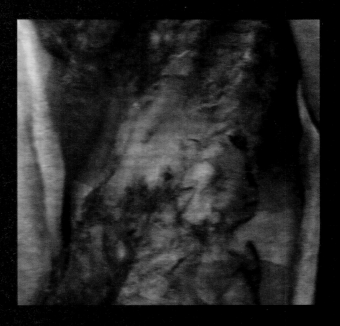

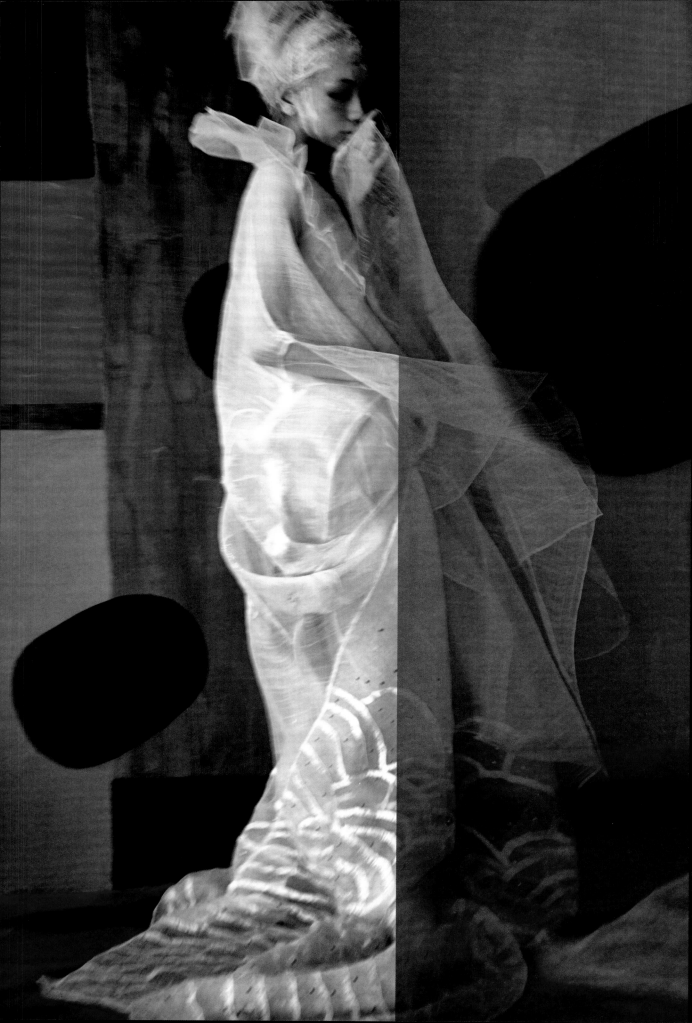

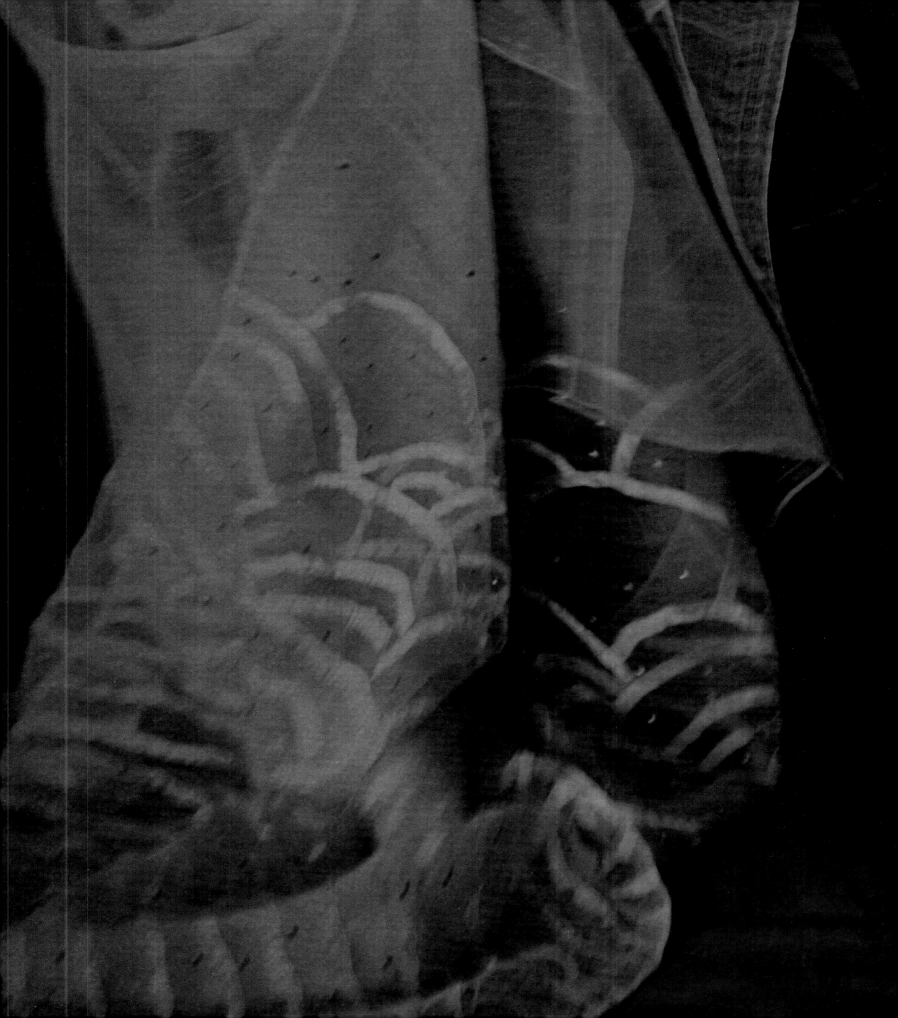

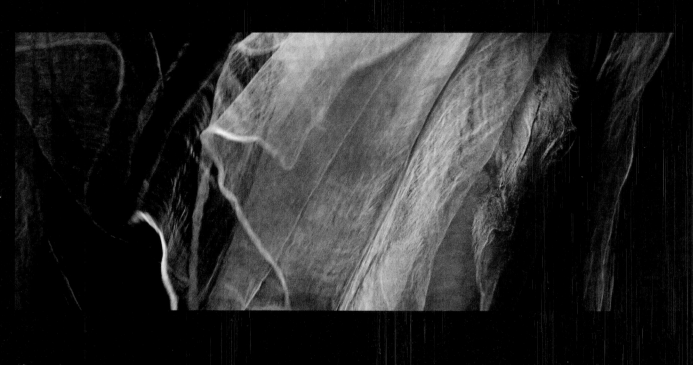

青绿山水

2009
绡、水纱
70cm×150cm

作品吸收了中国传统水墨画的韵味与意境，以黑、白色为主。在色彩自然的过渡与渐变中又呈现出一抹幽绿。线条自然流动，图案大气写意。结合柔软的丝质面料、简约大方的廓型，表达出清幽飘逸、自由随性的状态。

Turquois Landscape

2009
Silk
70cm×150cm

The work absorbs the charm and atmosphere of traditional Chinese ink painting. Mainly in black and white, in the gradual change and natural fading of color, it further shows a touch of green. The lines flow freely and naturally; the patterns are abstract and atmospheric. Combining soft silk fabrics with a simple and elegant silhouette, it expresses a free and ethereal state.

摄影：谷子
Photography: Guzi
化妆：君君
Make-up: Jun Jun
模特：王梓、张馨月、张珈琦
Models: Wang Zi, Zhang Xinyue, Zhang Jiaqi

072

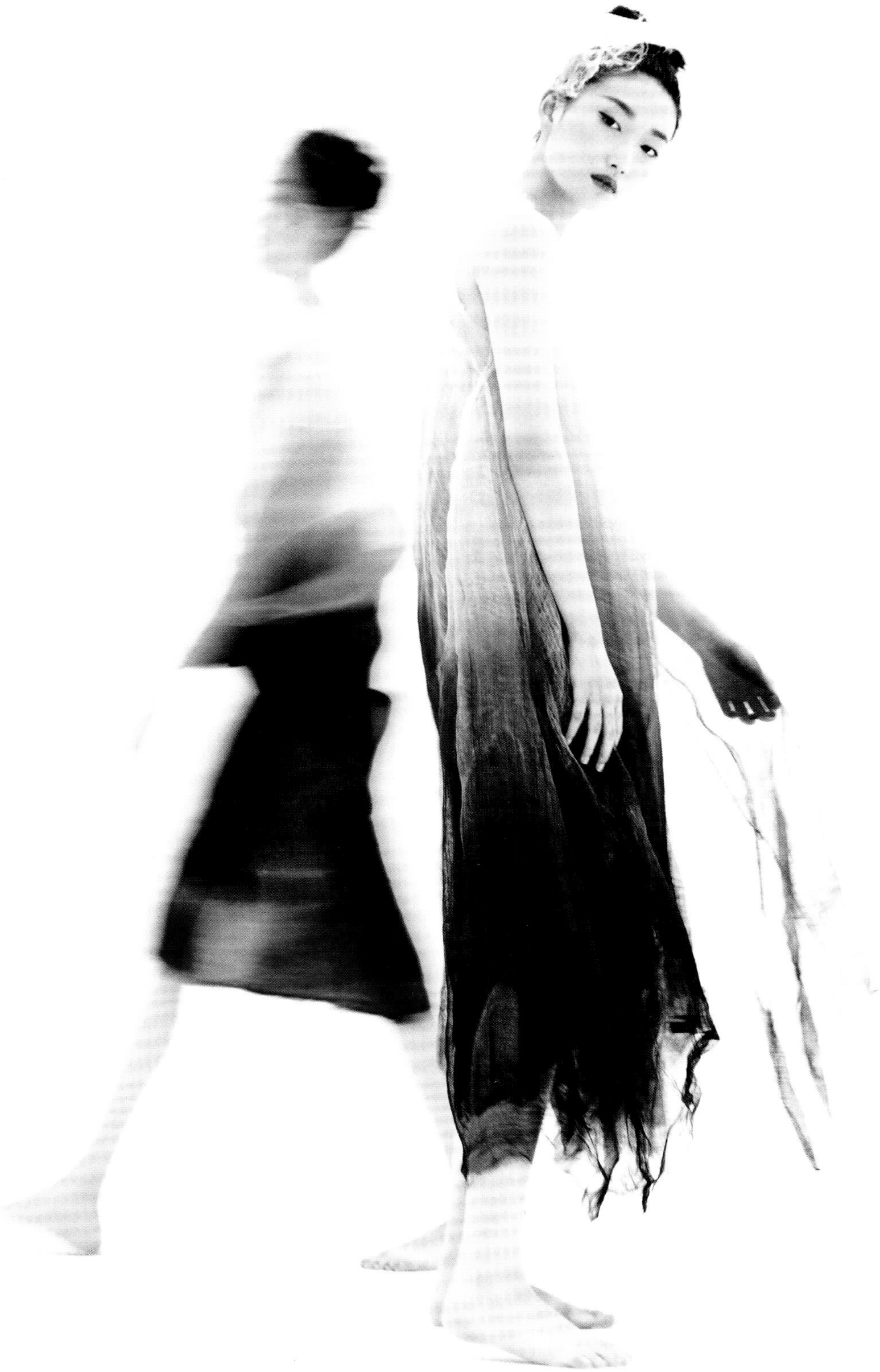

074

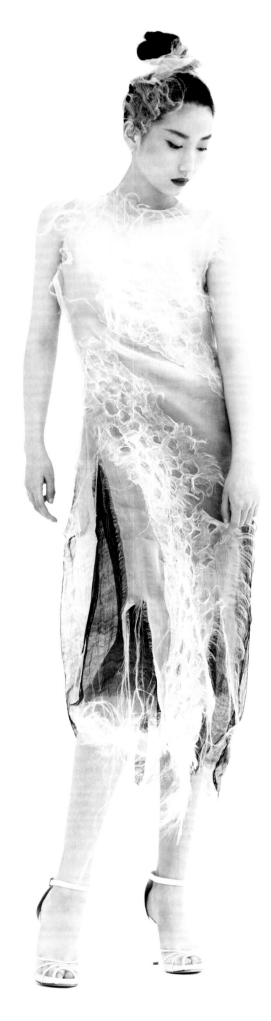

新中式

2011
绡、水纱
70cm×110cm

作品在民国旗袍的款制基础上,以更轻盈流动的方式重新解读传统。每件旗袍的款式与图案又各有特色,在小细节中实现视觉的协调与审美的享受,讲述传统与现代的关系与故事。

New Chinese Style

2011
Silk
70cm×110cm

Based on the style of cheongsam, this group of works reinterprets the tradition in a lighter and more fluid way. With a closer look, each cheongsam has its own style and pattern, which achieves visual coordination and aesthetic enjoyment in these small details.

摄影:谷子
Photography: Guzi
化妆:君君
Make-up: Jun Jun
模特:王梓、张馨月、张珈琦
Models: Wang Zi, Zhang Xinyue, Zhang Jiaqi

079

081

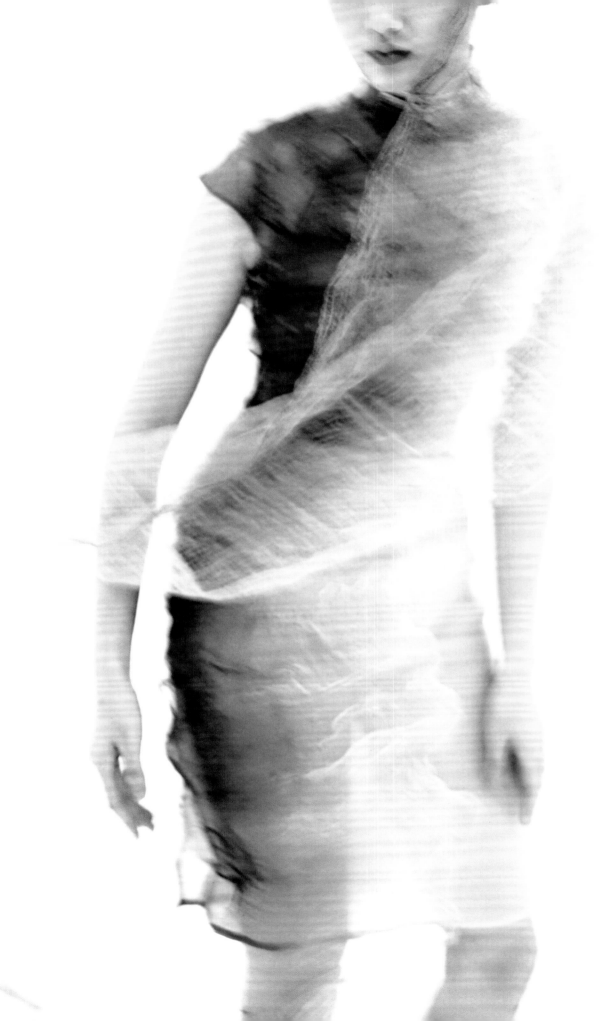

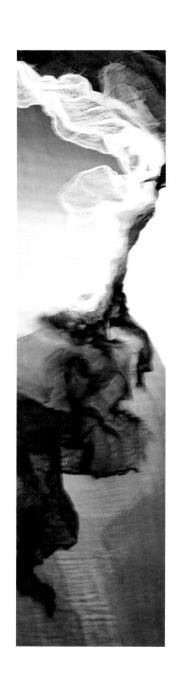

085

蓝之韵

2011
绡、水纱
70cm×110cm

  作品中的蓝色象征高雅与深远,经过蓝染的水纱,随着衣服的形制弯曲流动,江山水色青于蓝,传统的蓝穿越历史的长河呈现在当今的服装上,人、衣、色融为一体。

Rhyme of Blue

2011
Silk
70cm×110cm

  The use of blue in the work symbolizes elegance and far-reaching. The blue dyed yarn bends and flows with the shape of the clothes, making the river and the water blue in blue. The traditional blue crosses the long river of history and is presented on today's clothes, and people and clothes and colors are integrated.

摄影:谷子
Photography: Guzi
化妆:君君
Make-up: Jun Jun
模特:张珈琦
Model: Zhang Jiaqi

红色记忆

2011
绡、水纱
70cm×110cm

作品用简约时尚的形制承载着传统色彩的温度，经过层层浸染的红与黑，赋予绡和水纱新的生命力，一抹象征吉祥的中国红，凝结了中国数千年的文化内蕴，是历史与现代在时尚空间的连接。

Red Memory

2011
Silk
70cm×110cm

The series of works carries the temperature of traditional colors with a simple and fashionable shape. Through layers of dyed red and black, it endows the gauze and water yarn with new vitality. A touch of Chinese red, symbolizing auspice, condenses the cultural connotation of China for thousands of years, and is the connection between history and modernity in the fashion space.

摄影：谷子
Photography: Guzi
化妆：君君
Make-up: Jun Jun
模特：王梓、张馨月、张珈琦
Models: Wang Zi, Zhang Xinyue, Zhang Jiaqi

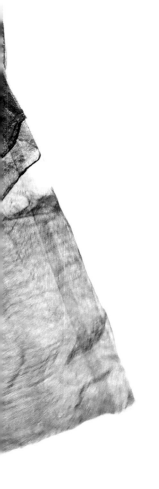

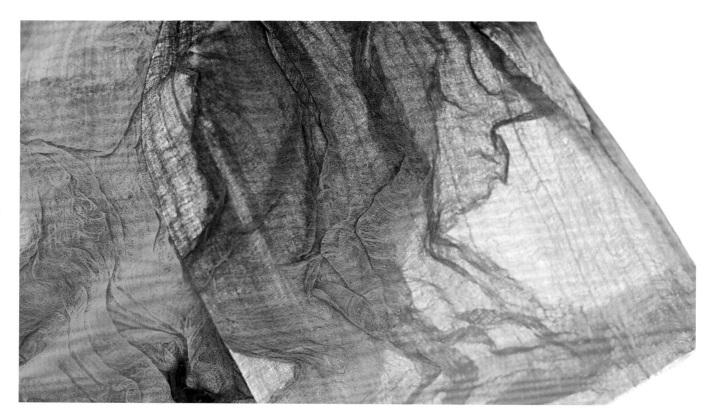

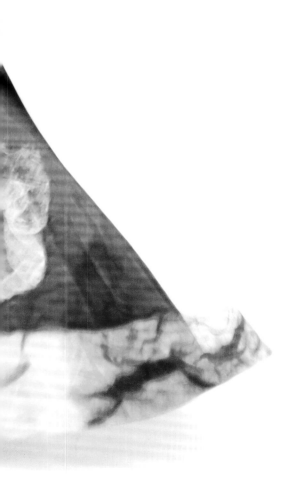

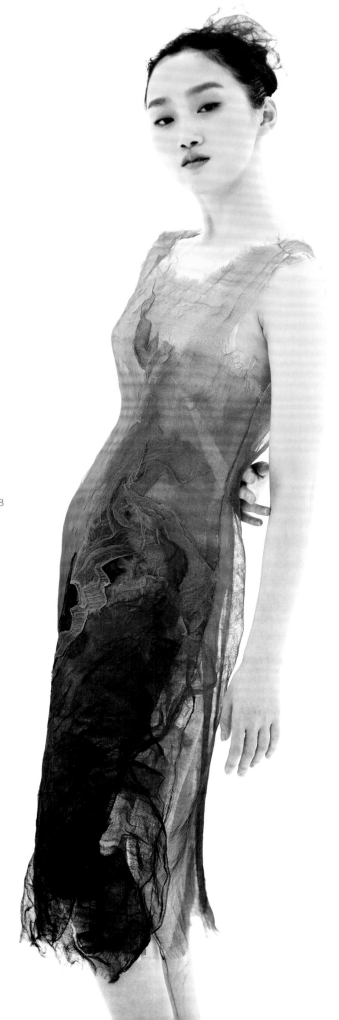

111

紫气东来

2011
真丝
70cm×110cm

作品用极具现代感的礼服形制展现独特的东方意蕴，绡、欧根纱等传统材质在当今科技与艺术融合下进行创新设计，飘动的紫色象征祥瑞，充满生气。

The Purple Spirit

2011
Silk
70cm×110cm

The series of works show a unique Oriental connotation with a very modern dress shape, and traditional materials such as gauze and organza as well as the innovative design of these materials under the integration of science and technology and art. The flowing purple symbolizes auspice and is full of vitality.

摄影：谷子
Photography: Guzi
化妆：君君
Make-up: Jun Jun
模特：王梓、张馨月、张珈琦
Models: Wang Zi, Zhang Xinyue, Zhang Jiaqi

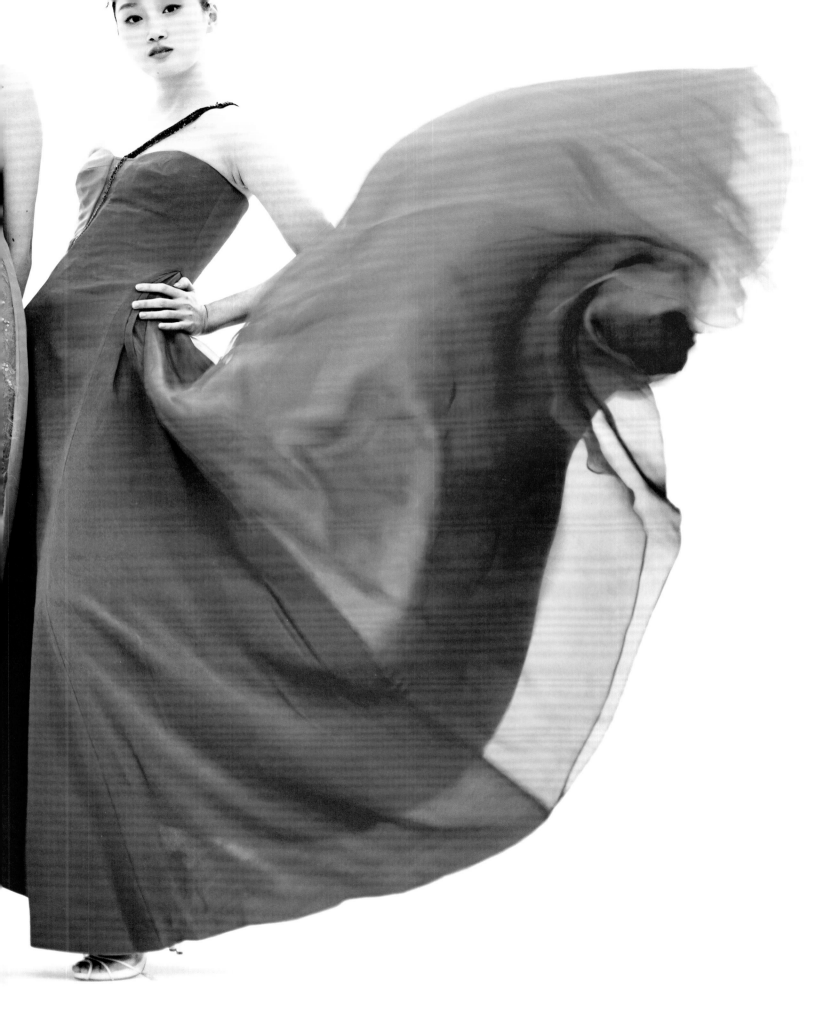

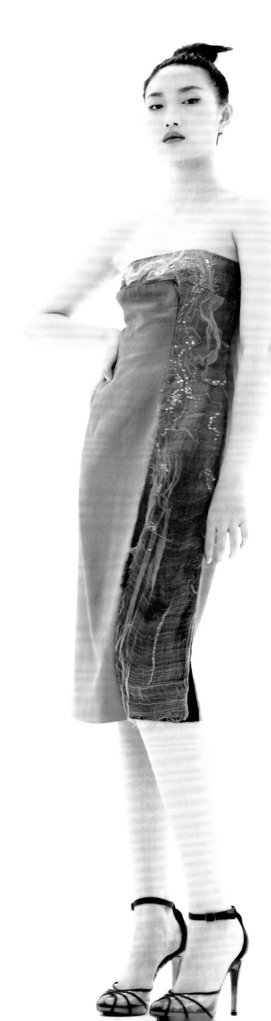

锦瑟华年

2022
宋锦

作品凸显了中国传统织锦与现代设计语言的融合,挖掘宋锦传统技艺的艺术价值及更丰富的表现形式,以促进宋锦服饰品艺术化、国际化的发展。"锦瑟无端五十弦,一弦一柱思华年"比喻青春年华,象征传统宋锦技艺焕发新的活力,也表达了宋锦服装设计的新理念及未来发展的多元化趋势。

Blooming Times

2022
Song brocade

The series highlights the fusion of traditional Chinese brocade and modern design language. It explores the artistic value of traditional Song brocade techniques and seeks richer forms of expression to promote the artistic and international development of Song brocade fashion and accessories. The poetic phrase by Li Shangyin compares the beauty of youth to the intricate strings of a harp, symbolizing the rejuvenation of traditional Song brocade techniques with new vitality. This also signifies the emergence of new concepts in Song brocade clothing design and reflects the diverse trends in its future development.

摄影:张润宇
Photography: Zhang Runyu
化妆:君君
Make-up: Jun Jun
模特:张宁
Model: Zhang Ning

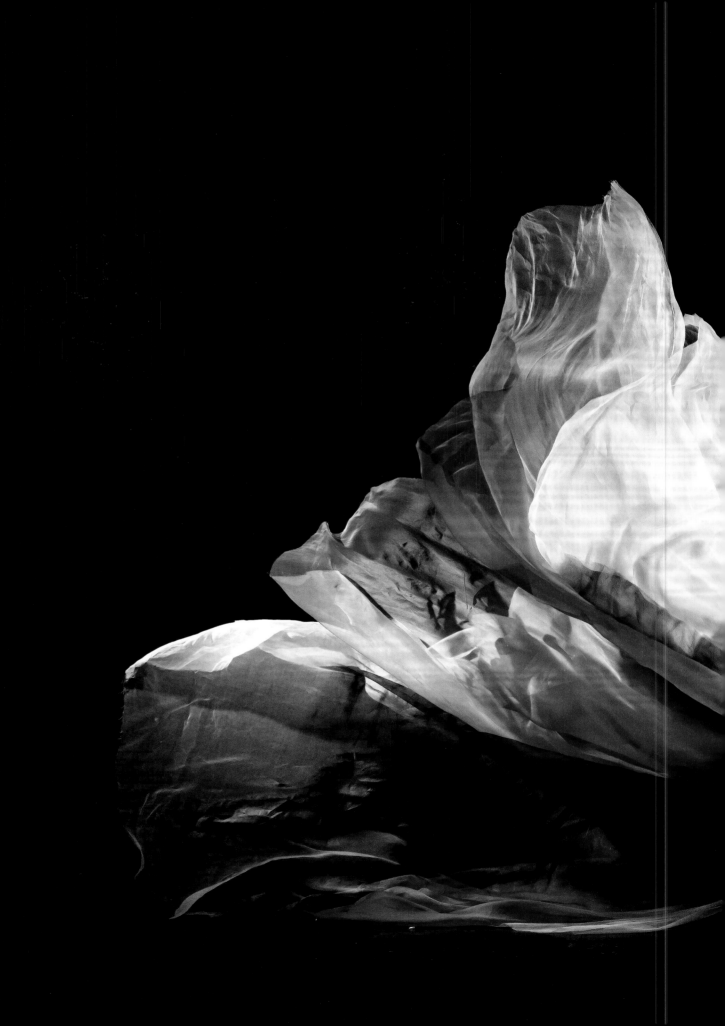

摄影：谷子
Photography: Guzi
化妆：君君
Make-up: Jun Jun
模特：张珈琦
Model: Zhang Jiaqi

139

荷塘月色

2018
缎、苏绣
60cm×110cm

作品灵感来源于朱自清《荷塘月色》一文中描绘的意境。水墨韵味的刺绣线条柔润而富有变化，浓淡相间的荷塘铅华洗尽，服装整体效果清新淡雅，散发着江南情韵。

Moonlight over the Lotus Pond

2018
Satin, Su embroidery
60cm×110cm

The inspiration comes from the artistic conception depicted in Zhu Ziqing's *Moonlight over the Lotus Pond*. The embroider line with lasting appeal of water and ink is silky and moist and be full of change, the lead of the lotus pond of thick and thin alternate with is washed completely, garment whole effect is pure and fresh quietly elegant, sending out the amorous feelings of south of the Yangtze river.

摄影：谷子
Photography: Guzi
化妆：君君
Make-up: Jun Jun
模特：张馨月、张珈琦
Models: Zhang Xinyue, Zhang Jiaqi

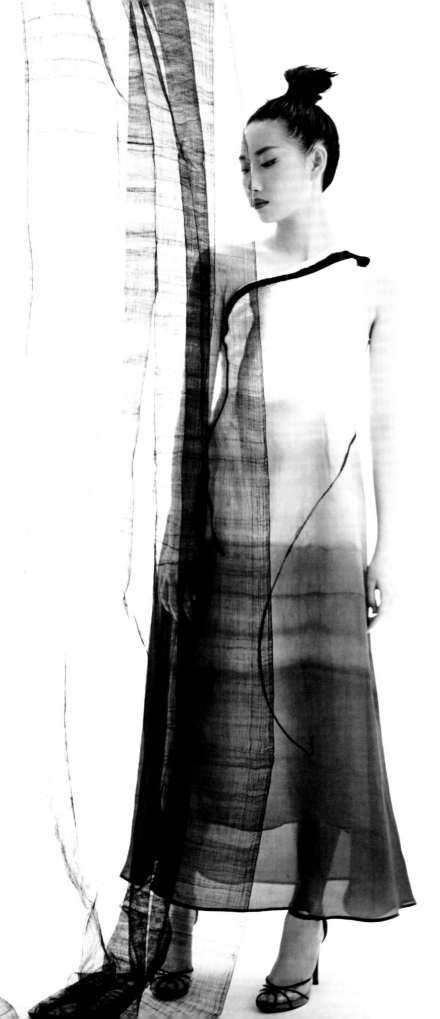

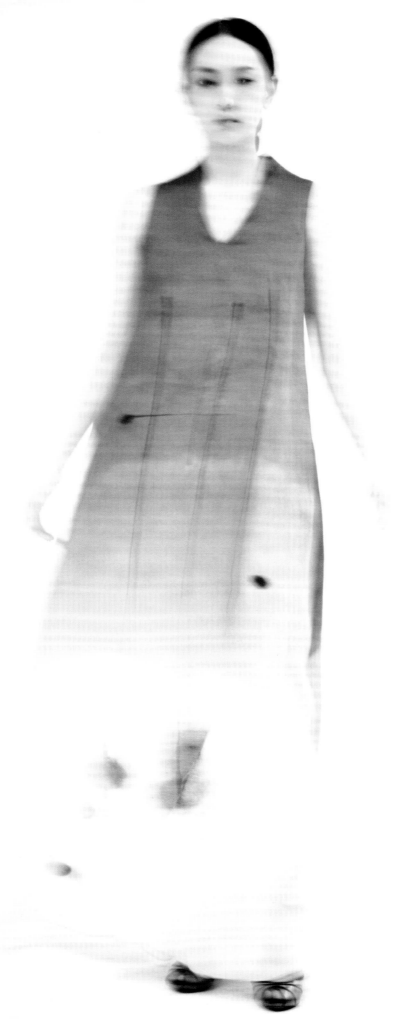

145

雅韵

2018
缎、苏绣
60cm×110cm

作品从中国历代龙袍中提取龙凤、云水等吉祥纹样的精华，在追寻传统东方视觉审美的同时，将其与现代美学、当代设计语言相融合，把传统刺绣手法用现代语汇进行表达。

Elegance

2018
Satin, Su embroidery
60cm×110cm

The essence of auspicious patterns such as dragon phoenix and cloud water is extracted from the dragon robe in the previous dynasties of the style. While pursuing the traditional Oriental visual aesthetics, the style is integrated with modern aesthetics and contemporary design language, and the traditional embroidery technique is expressed in modern vocabulary.

摄影：张润宇
Photography: Zhang Runyu
化妆：君君
Make-up: Jun Jun
模特：张宁
Model: Zhang Ning

147

150

花梦敦煌

2018
丝，宋锦
60cm×110cm

千里一梦，敦煌花开。作品灵感取自莫高窟壁画中的花卉纹样，结合服装形态变幻重组，使敦煌之花绽放当代之美。

Flower Dream of Dunhuang

2018
Silk, Su embroiledry
60cm×110cm

Inspired by the flower patterns in the murals of Mogao grottoes and combined with the changing and recombination of clothing forms, the flowers of Dunhuang bloom with contemporary beauty.

摄影：谷子
Photography: Guzi
化妆：君君
Make-up: Jun Jun
模特：王梓、张馨月、张珈琦
Models: Wang Zi, Zhang Xinyue, Zhang Jiaqi

161

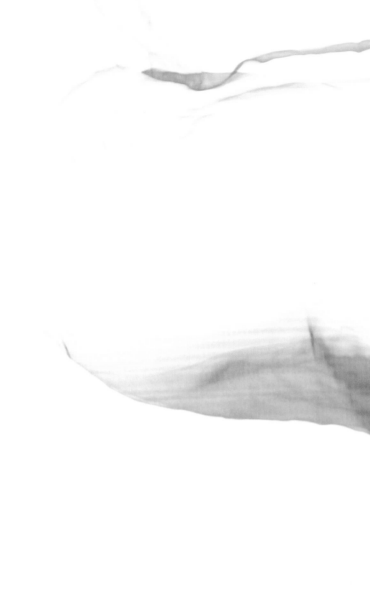

163

165

空与影

2017
黑色弹性 TPU 材料
150cmX70cm

作品通过 3D 打印实现衣与人之间由二维及三维的交融，由形状到体块的层层叠叠，具象实质的衣和抽象虚幻的影呈现虚虚实实的空间，既表达出作品灵动的意境，又表现了设计师的人格风度及个性情感。

Emptiness and Shadow

2017
Black elastic TPU material
150cmX70cm

The work achieves a fusion between 2D and 3D spaces between clothing and the human body through 3D printing. Layer by layer, it transitions from shapes to forms, presenting a juxtaposition of tangible clothing and abstract, ethereal shadows. This interplay of real and illusory spaces conveys the dynamic artistic concept of the work, while also reflecting the designer's personality, style, and individual emotions.

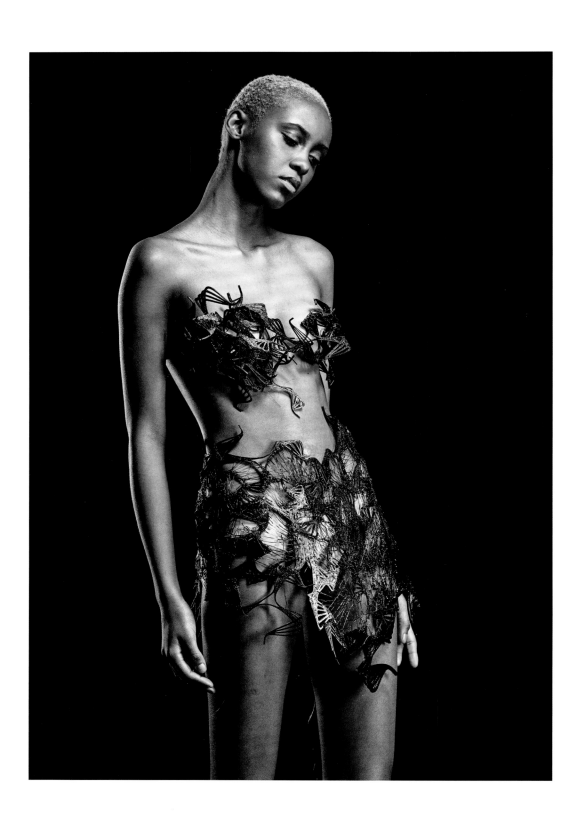

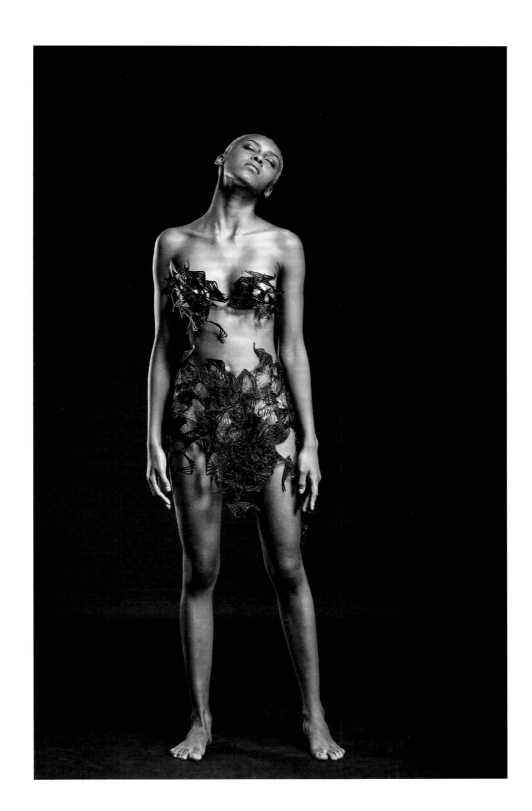

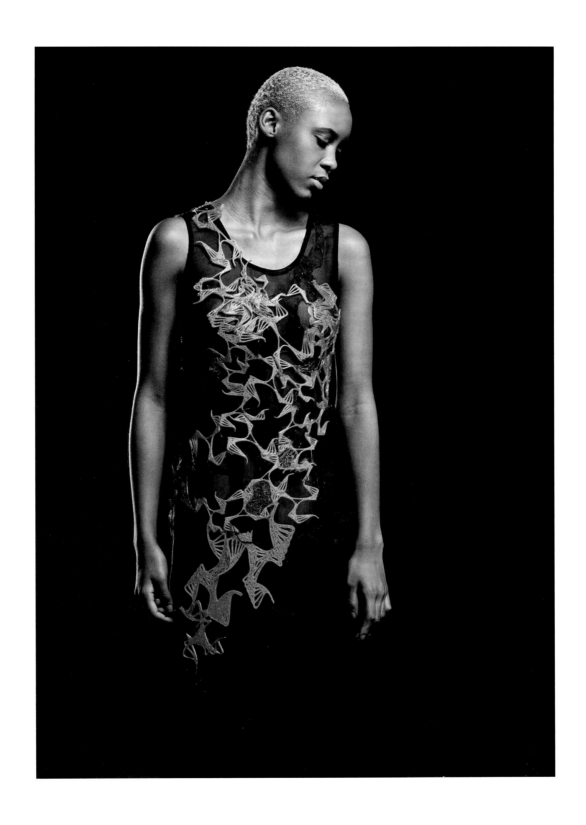

171

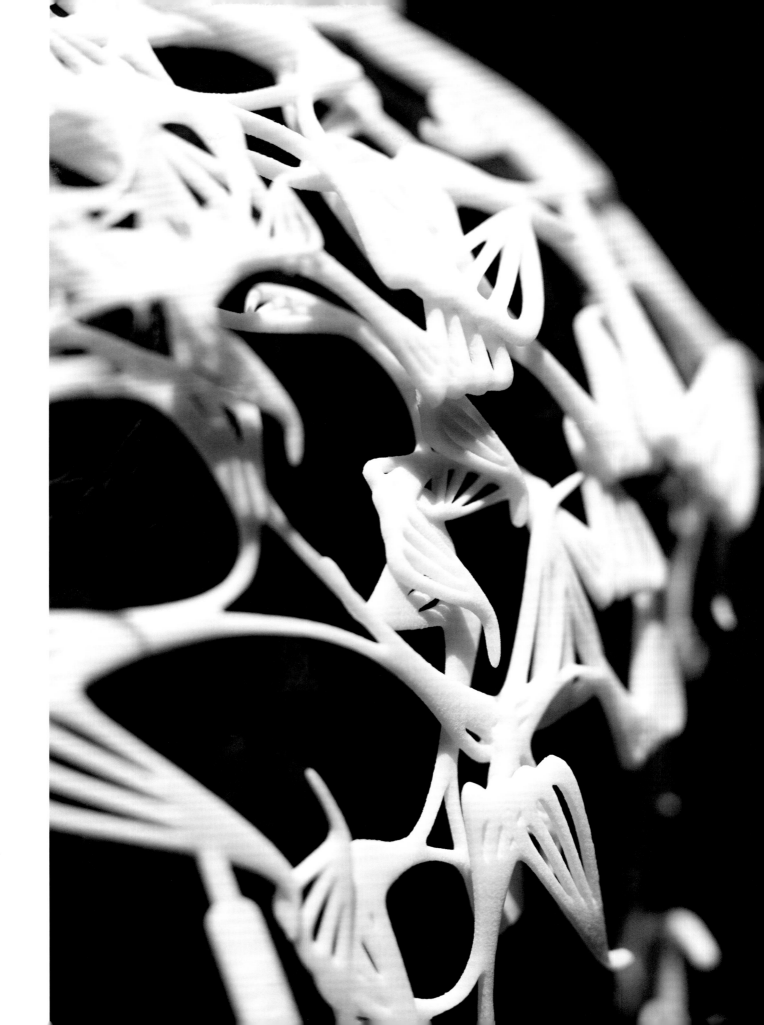

述说

2009
亚克力
60cm×90cm

　　作品运用方形硬质亚克力连缀成小礼服造型，拼接结构形似兵马俑盔甲。透过时空隧道，以新的设计表达连接古与今的光影，赋予象征吉祥、喜庆的中国红以新的意义，这是远古文化图腾与当今观者的连通与对话。

Narration

2009
Acrylic
60cm×90cm

　　The artwork employs square, rigid acrylic pieces interconnected to form the shape of a mini dress, the structure of which resembles the armor of Terracotta Warriors. Through a tunnel of time and space, it employs innovative design to convey the interplay of light and shadow that bridges the ancient and the contemporary. This grants a fresh significance to the symbolically auspicious and festive Chinese red. It represents a connection and dialogue between ancient cultural totems and contemporary observers.

清·远·静

2010
绡、水纱
110cm×800cm

作品意在以丝帛呈现中国山水画的艺术形式，表现水墨画的灵动气韵和空间感。穿梭在壁挂中，空阔悠远的浮动感令人不禁思考时间和生命的流逝……吊染工艺和水纱面料的融合呈现出丰富的视觉效果，似云非云、似山非山，表达"江流天地外，山色有无中"的清远意境。

Qing Yuan Jing

2010
Gauze
110cm×800cm

This series of artworks aims to present the artistic essence of traditional Chinese landscape painting using silk fabric, capturing the dynamic rhythm and spatial depth of ink and wash painting. As one moves within the hanging pieces, the expansive and distant floating sensation prompts contemplation of the passage of time and life. The fusion of the dip dyeing technique and aqua fabric creates a rich visual effect, portraying images that seem like clouds and mountains, yet not entirely. This conveys the serene and distant artistic conception of "beyond the river, beyond the sky; between the mountains, existence and nonexistence."

185

行云流水

2006

材料：真丝绡、水纱、光导纤维、LED
工艺：LED 链接光导纤维，光导纤维随着音乐节奏的变化闪动，声音越大，LED 越亮，反之越暗，是丝与声、光、电的完美融合。
360cm×210cm

作品强调艺术与科学的多元融合。屏风中，水纱与绡的层层透叠表达江南水乡的婉约气质，结合声、光、电多种元素呈现中国水墨画的意境。万千的光点随着音乐节奏的起伏在水纱与光导纤维中自由穿梭，似风中云雾升腾起落，形成微妙的变化，在时而铿锵、时而舒缓的旋律中体验丝、声、光、电的融合互动。

Flowing Clouds, Running Water

2006

Materials: Organza, Aqua Fabric, Optical Fibers, LED

Techniques: The artwork involves linking LEDs to optical fibers, where the optical fibers respond to changes in the rhythm of the accompanying music. As the sound increases in intensity, the LEDs shine brighter, and conversely, they dim in response to softer sounds. This synchronization of silk with sound, light, and electricity creates a harmonious fusion.

360cm×210cm

The artwork emphasizes the diverse fusion of art and science. Within the screen, layers of aqua fabric and silk express the gentle and elegant qualities of the Jiangnan water village. Combined with elements of sound, light, and electricity, it presents the artistic conception of Chinese ink and wash painting. Countless points of light freely traverse through the aqua fabric and optical fibers in sync with the rhythm of music, resembling the rising and falling of mist in the wind. These create subtle changes, allowing one to experience the harmonious interaction of silk, sound, light, and electricity within the ever-changing melody of both strength and calmness.

风声

2019

材料：真丝绡、水纱、光导纤维、LED
工艺：LED连接光导纤维，光导纤维随着音乐节奏的变化闪动，声音越大，LED越亮，反之越暗，是丝与声、光、电的完美融合。
300cm×210cm

作品通过形塑真丝绡、水纱的肌理，结合声、光、电等多种元素呈现中国水墨的磅礴气势。屏风中的光点随着音乐节奏的起伏在水纱与光导纤维中自由穿梭，如同北方旷野的风，在群山中呼啸，形成无穷的变化。在冲撞与缓行、高歌与低吟中达到视听的共鸣。

Sound of the Wind

2019

Materials: Organza, Aqua Fabric, Optical Fibers, LED
Techniques: The artwork involves linking LEDs to optical fibers, where the optical fibers respond to changes in the rhythm of the accompanying music. As the sound increases in intensity, the LEDs shine brighter, and conversely, they dim in response to softer sounds. This synchronization of silk with sound, light, and electricity creates a harmonious fusion.
300cm×210cm

The artwork achieves a majestic representation of Chinese ink painting by shaping the texture of silk organza and aqua fabric, combined with various elements such as sound, light, and electricity. Within the screen, points of light move freely through aqua fabric and optical fibers in response to the rhythm of the music, much like the wind howling across the vast northern wilderness, creating endless variations. It attains a visual and auditory resonance through the collision and gentle flow, as well as the high and low tones, immersing the viewer in its captivating presentation.

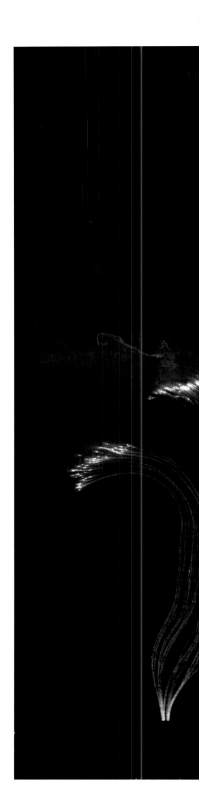

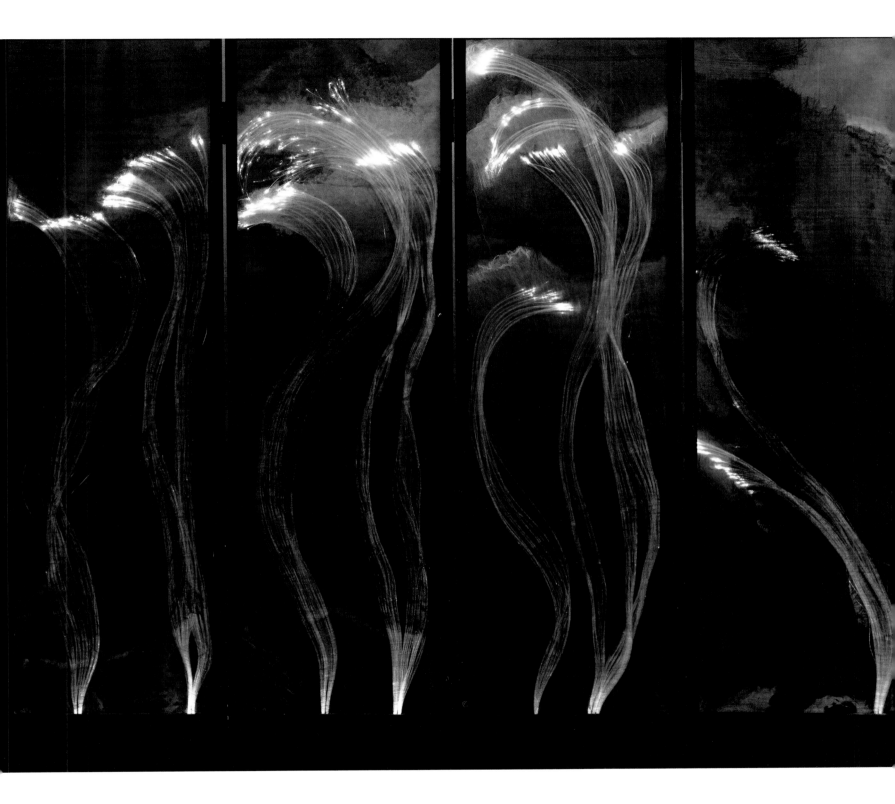

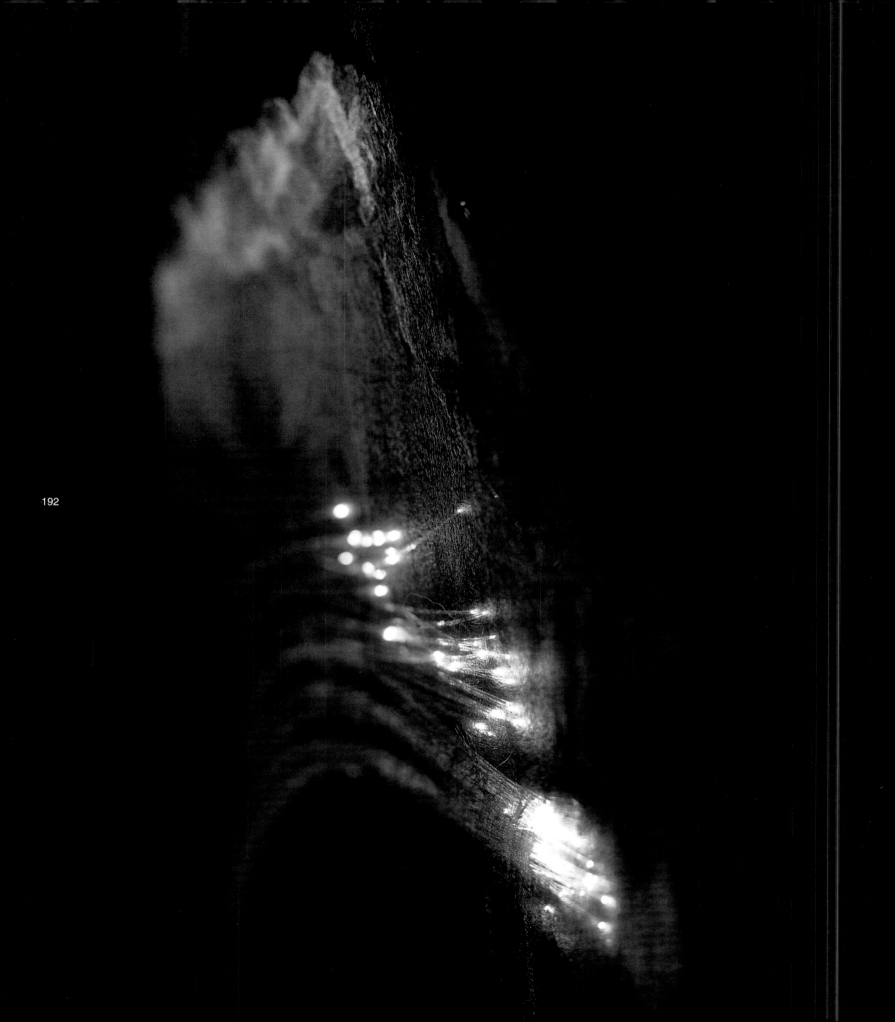

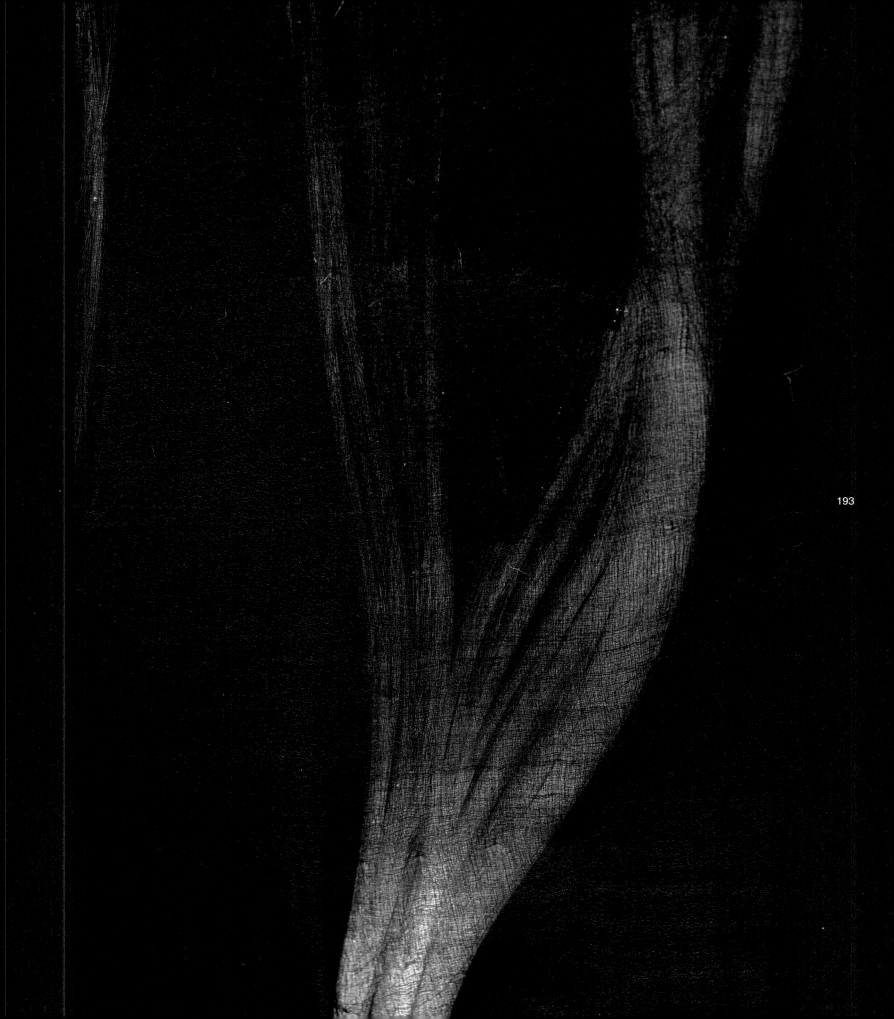

夜与昼·丝网版画

2022
80cm×115cm

　　作品旨在用丝网版画的语言再现服装艺术作品《夜与昼》的神韵与精髓。作品通过自然底色、金底和银底的厚印法细腻而生动地重现了这件服装,这三种底色的作品分别显示出质朴、华贵和内敛的气息,整幅画面浑然天成。

Night and Day: Silkscreen Prints

2022
80cm×115cm

　　This work aims to capture the essence and charm of the fashion artwork "Night and Day" using the language of silkscreen printing. Through the use of natural background color, gold, and silver base printing techniques, this work vividly reproduces the original piece. Each of these three base colors conveys a sense of simplicity, luxury, and restraint, creating a harmonious and natural composition in the artwork.

山峦

2019
真丝绡、水纱
300cm×30cm

作品刻画了巍峨的群山，一抹中国红与水墨山水意趣相结合，红色水纱似云雾般升腾，形成无穷的变化，两种元素在矛盾与冲撞中达到了一种平衡。

Mountain Range

2019
Gauze
300cm×30cm

This work depicts towering mountains, where a touch of Chinese red merges with the charm of an ink-and-wash landscape. The red aqua fabric rises like mist, creating infinite variations. The interplay between these two elements achieves a delicate balance amid their contrast and collision.

枯山水

2022
夏布
90cm×1300cm

　　作品将夏布进行热压塑形处理，在有形与无形，具象与抽象之间，营造山岩的肌理效果，层叠的褶皱在转折起伏中构筑成有节奏的山体空间，既表现群山的厚重，又表达出智者乐水，仁者乐山的感悟。

Dry Landscape

2022
Grasscloth
90cm×1300cm

    The artwork undergoes a heat-pressed shaping process using grasscloth. In the realm between tangible and intangible, concrete and abstract, it crafts the texture of rocky mountains. The layered folds, in their undulating transitions, construct a rhythmic spatial composition of mountain forms. This presentation captures the weightiness of mountain ranges while also conveying the insight that wise individuals find joy in water, and kind-hearted ones find delight in mountains.

2022
绡、水纱
90cm×1300cm

作品运用传统植物染色，将绡与水纱通过特殊工艺处理形成辽阔空旷的丝绸之路，在漫漫丝路上记载着中国传统文化的传播与交流，是丝路印象的静态表达也是文化基因的艺术再现。

Smoke in the Desert

2022
Gauze
90cm×1300cm

The artwork utilizes traditional plant dyeing techniques to transform raw silk and aqua fabric through special processes, creating a vast and open Silk Road. It records the dissemination and exchange of Chinese traditional culture along the long Silk Road. This is not only a static expression of Silk Road impressions but also an artistic representation of cultural heritage.

乐山水

2022
绡、水纱、绢
130cm×750cm

　　作品采用中国传统材质绢，运用白族手工吊染工艺结合手绘形成近山远山之间丰富的层次关系；以植物染绡、绢、水纱及夏布等材质表现山川层峦广阔悠远、大气磅礴的气势意境，是对中国山水印象的静态表达，也是对传统文化基因的艺术再现。以不同形式的装置作陈列布局，在转折起伏中构筑成有节奏的山水空间，观者穿梭其中沉浸式体验与山水的连通。精神所到，气韵以生，超乎形状迹象之外，在山水之间见天地见自我。

Joyful Landscape

2022
Gauze, Silk
130cm×750cm

　　This series of installation artworks employ traditional Chinese material, silk, using the handmade tie-dye technique of the Bai ethnic group combined with hand-painting to create intricate layering effects between near and distant mountains. Through the use of plant-dyed silk, raw silk, aqua fabric, and grasscloth, it portrays the grandeur and vastness of mountains and rivers, capturing an imposing and expansive artistic conception. This serves as a static expression of the impressions of traditional Chinese landscape art and a reimagining of the cultural roots. Using diverse installation formats and arrangements, these pieces construct a rhythmic landscape space in their undulating transitions, allowing viewers to immerse themselves and experience a connection with the landscape. As the viewer engages, the vitality comes to life, transcending mere forms and traces, and amidst the mountains and waters, they witness the meeting of heaven, earth, and self.

大事记

服装秀

《李薇艺术服装秀》，东京中国文化中心，2018

《李薇3D打印服装秀》，2018国际3D打印嘉年华，2018

《李薇艺术服装秀》，肯尼亚内罗毕大学，马赛部落，2018

《红之韵》，首尔时尚艺术节，韩国首尔，2018

《Xiuniang·李薇高级定制时装发布会》，北京798"79立方"，2014

《"一脉相承"李薇师生作品发布》，宁波时装周，2014

《李薇高级定制时装发布会》，大连服装周Z28，2016

《李薇高级定制服装秀》，新疆国际会展中心，2017

《李薇3D打印服装秀》，北京紫云轩，2017

《绣在江南——李薇作品发布》，苏州中国刺绣文化艺术节开幕式，2017

《李薇艺术服装秀》，华润集团总部深圳春笋大厦，2018

《李薇艺术服装秀》，丝绸之路国际时装周，广州南沙，2019

《纤墨窈然——李薇艺术服装秀》，锦绣中华非遗服装秀，景山公园，2019

《李薇艺术服装秀》，绣美中国文化沙龙，北京天桥演艺中心，2019

《七彩云南》，"丝路云裳·七彩云南2021昆明民族时装周"，云南昆明，2021

《锦瑟华年——李薇宋锦服装秀》，2021AFC亚洲时尚品牌盛典，四川成都，2021

《人·见本心——丝路云裳·穿在身上的艺术（第三季）》，云南卫视，2021

《新中式》，第五届沈阳旗袍文化节旗袍服装秀，辽宁沈阳，2021

《锦瑟年华——上久楷·李薇服装秀》，2021

《南溟吉贝——黎锦主题非遗服饰秀》，锦绣中华——2021中国非物质文化遗产服饰秀，海南三亚，2021

《简约的艺术——丝路云裳·穿在身上的艺术（第二季）》，云南卫视，2021

Chronicle of Events

Fashion show

" Li Wei Art Fashion Show", China Cultural Center, Tokyo, 2018

" Li Wei 3D Printing Fashion Show", International 3D Printing Carnival, 2018

" Li Wei Art Fashion Show", University of Nairobi in Maasai Tribe, Kenya, 2018

" Rhyme of Red", Seoul Fashion Week, Seoul, South Korea, 2018

" The Xiuniang • Li Wei Haute Couture Fashion Work Show", Beijing 798 Art Zone, 2014

" The Continuation: Li Wei Teacher-Student Work Show", Ningbo Fashion Week, 2014

" Li Wei Haute Couture Fashion Show", Dalian Fashion Week Z28, 2016

" Li Wei Haute Couture Fashion Show", Xinjiang International Convention and Exhibition Center, 2017

" 3D Printed Fashion Show of Li Wei", Beijing Ziyunxuan, 2017

" Embroidered in Jiangnan - Li Wei Work Show", China Embroidery Art Festival Opening Ceremony, Suzhou, 2017

" Li Wei Art Fashion Show", China Resources Headquarters, Shenzhen Chunshun Building, 2018

" Li Wei Art Fashion Show", Silk Road International Fashion Week, Guangzhou Nansha, 2019

" Ink and Beauty - Li Wei Art Fashion Show", China Intangible Cultural Heritage Fashion Show, Jingshan Park, 2019

" Li Wei Art Fashion Show", Splendid China Cultural Salon, Beijing Tianqiao Acrobatics Theatre, 2019

" Colorful Yunnan," "Silk Road, Colorful Yunnan 2021 Kunming Ethnic Fashion Week", Kunming, Yunnan, 2021

" Melody of Brocade: Li Wei Song Brocade Fashion Show", 2021 AFC Asian Fashion Brand Ceremony, Chengdu, Sichuan, 2021

" Human True Self: Silk Road, Art on the Body (Season 3)", YNTV, 2021

" New Chinese Style", The 5th Shenyang Cheongsam Culture Festival Cheongsam Fashion Show, Shenyang, Liaoning, 2021

" Colorful Youth: Shangjiu Kai, Li Wei Fashion Show", 2021

" Nanming Jibe: Li Jin themed Intangible Cultural Heritage Costume Show", Splendid China - 2021 Chinese Intangible Cultural Heritage Costume Show, Sanya, Hainan, 2021

" Simplicity in Art: Silk Road, Art on the Body (Season 2)", YNTV, 2021

展览

"李薇中国水墨画展",法国巴黎艺术城画廊,1995

"李薇服装艺术展",法国巴黎艺术城画廊,2002

"李薇中国京剧脸谱画展",法国巴黎印象画廊,2003

"李薇艺术作品展",英国皇家艺术学院,2016

"视·听李薇艺术作品展",意大利米开朗琪罗·皮斯多莱托艺术中心,2013

"李薇艺术作品展",798悦美术馆,2014

"李薇艺术作品展",香港理工大学,2017

"夜与昼——李薇艺术作品展","北京国际设计周",北京农业展览馆,2019

"锦瑟华年——李薇宋锦服装艺术展","黄河非遗国际创意周",河南洛阳,2021

《乐山水》,"BMW中国文化之旅非遗保护创新成果展",北京中国科学技术馆,清华大学美术学院、中国科技馆、宝马中国联合主办,2022

"十国艺术家联展",法国巴黎电台,2002

"身体中国"中国当代艺术展,法国马赛当代艺术馆,2004

"中国艺术展览",欧洲商学院,2004

"国际设计双年展",韩国光州金大中美术馆,2005

西班牙马德里文化中心,2012

"中国当代纤维艺术展",西班牙哥伦比亚圣多明戈大剧院、巴西圣保罗"卡米"艺术中心、厄瓜多尔、乌克兰,2012

"超越历史与物质:中国当代丝绸艺术展",德国柏林中国文化艺术中心、西班牙文化,2013

"时尚艺术双年展",韩国首尔弘益艺术中心,2016

"波兰洛兹设计双年展",波兰洛兹纤维艺术博物馆,2016

"亚洲纤维艺术展",日本福冈Koryu画廊,2017

"丝路映像——中国时装艺术精品展",东京中国文化中心,2018

"超越表面——姜绶祥及其合作设计师布与衣的探索",美国康奈尔大学,2018

Art Exhibition

" Li Wei Chinese Ink Painting Exhibition", Cité Internationale des Arts, Paris, France, 1995

" Li Wei Fashion Art Exhibition", Cité Internationale des Arts, Paris, France, 2002

" Li Wei Chinese Opera Facial Makeup Painting Exhibition", Galerie Impressions, Paris, France, 2003

" Li Wei Art Exhibition", Royal Academy of Arts, United Kingdom, 2016

" Sight Sound Li Wei Art Exhibition", Italy-Michelangelo Pistoletto Art Center, 2013

" Li Wei Art Exhibition", 798 Art Gallery, China, 2014

" Li Wei Art Exhibition", Hong Kong Polytechnic University, Hong Kong, 2017

" Night and Day, Li Wei Art Exhibition", "Beijing International Design Week", China Agriculture Exhibition Center, Beijing, 2019

" Colorful Youth, Li Wei Song Brocade Costume Art Exhibition", "Yellow River Intangible Cultural Heritage - International Creative Week", Luoyang, Henan, China, 2021

" Le Shan Shui", "BMW China Culture Tour - Intangible Cultural Heritage Innovation Exhibition", China Science and Technology Museum, Tsinghua University Academy of Fine Arts, jointly hosted by China Science and Technology Museum and BMW China, 2022

" Joint Exhibition by Artists from Ten Countries", RFI, France, 2002

" Body", China Musée d'Art Contemporain Marseille, France, 2004

" Chinese Art Exhibition", European Business School, France, 2004

" International Design Biennale", Kim Dae Jung Convention Center, Gwangju, South Korea, 2005

Art Center of Madrid Cultural Center, Spain, 2012

" Chinese Contemporary Fiber Art Exhibition", Teatro Mayor Julio Mario Santo Domingo, The São Paulo Museum of Art, Ecuador, Brazil, and Ukraine, 2012

" Beyond History and Material: Chinese Contemporary Silk Art Exhibition", China Cultural Art Center and Spain Cultural Art Center, Berlin, Germany, 2013

" Fashion Art Biennale", Art Center of Seoul Hongik University, South Korea, 2016

" Poland Łódź Design Biennale", Center Museum of Textiles, Łódź, Poland, 2016

"首尔时尚艺术节"，首尔BEAT360，2018

"中国纤维艺术世界巡展——意大利展"，佛罗伦萨普拉托纺织博物馆，2018

"设计中国·丝路花语参展"，吉尔吉斯斯坦，2019

"丝路映像——中国时装艺术精品展"，毛里求斯、俄罗斯、德国，2019

"2020 国际时装艺术双年展"，韩国首尔，2020

"2022 国际时装艺术双年展"，韩国釜山，2022

第二届AIAC ARTS"相遇当代艺术"国际当代艺术展，法国里昂市，法国国际当代艺术协会，2022

"艺术与科学国际作品展"，中国美术馆，2001

"第十届全国美术展览"，中国美术馆，2004

"第十一届全国美术展览"，中国美术馆，2009

"从洛桑到北京"第六届国际艺术双年展，河南美术馆，2010

"时尚脚本——纤维美学"七国艺术家联展，香港理工大学，2011

"首届北京国际设计三年展"，首都博物馆，2011

"艺术与科学国际作品展"，中国美术馆，2011

"第三届艺术与科学国际作品展"，中国科技馆，2012

"FAISION ART 服装艺术展"，爱慕美术馆，2013

"第十二届全国美术展览"，西安美术馆，2014

"FAISION ART 时装艺术国际展"，筑中美术馆，2017

"首届'一带一路'非中艺术交流展暨非中文化论坛"，中国国家图书馆，2018

"从洛桑到北京"第十届国际纤维艺术双年展，清华大学艺术博物馆，2018

"丝路艺蕴——中欧女性艺术交流展"，北京恭王府博物馆，2018

"不同的声音"纤维艺术的关怀叙事——10国艺术家联展，清华大学美术学院美术馆，2018

"第十三届全国美术展览"，山东工艺美院美术馆，2019

"第五届艺术与科学国际作品展"，中国国家博物馆，2019

"交织的视野——中美纤维艺术与科技创新交流展"，清华大学美术学院、美国费城艺术大学，2020

2021年"艺科融合·质创未来"系列主题展览，清华青岛艺术与科学创新研究院，2021

"同构——当代艺术作品邀请展"，深圳罗湖美术馆，2021

"云荟：中国时尚回顾大展2011-2020"，中国丝绸博物馆，2021

" Asian Fiber Art Exhibition", Koryu Museum, Fukuoka, Japan, 2017

" Silk Road Splendid, Chinese Fashion Art Boutique Exhibition", China Cultural Center, Tokyo, Japan, 2018

" Beyond Surface, Exploring Fabric and Clothing of Jiang Shouxiang and His Collaborative Designers", Cornell University, United States, 2018

" Seoul Fashion Art Festival", Seoul BEAT360, South Korea, 2018

" China Fiber Art World Tour Exhibition - Italy", Florence Prato Textile Museum, Italy, 2018

" Design China, Silk Road Exhibition", Kyrgyzstan, 2019

" Silk Road Splendid, Chinese Fashion Art Boutique Exhibition", Mauritius, Russia, Germany, 2019

" 2020 International Fashion Art Biennale", Seoul, South Korea, 2020

" 2022 International Fashion Art Biennale in Busan", Busan, South Korea, 2022

AIAC ARTS, Exposition Rencontres Contemporaines, Lyon, France, 2022

" Art and Science International Exhibition and Symposium", National Art Museum of China, 2001

" The 10th National Art Exhibition", National Art Museum of China, 2004

" The 11th National Art Exhibition", National Art Museum of China, 2009

"From Lausanne to Beijing", the 6th International Fiber Art Biennale Exhibition, Henan Art Museum, 2010

" Fashion Script, Fiber Aesthetics", Joint Exhibition, Hong Kong Polytechnic University, 2011

" The First Beijing International Design Triennial", National Museum of China, 2011

" Art and Science International Exhibition and Symposium", National Art Museum of China, 2011

" The 3rd Art and Science International Exhibition and Symposium", China Science and Technology Museum, 2012

" FAISION ART Fashion Art Exhibition", Aimu Art Museum, 2013

" The 12th National Art Exhibition", Xi'an Art Museum, 2014

" FAISION ART Fashion Art International Exhibition", Zhuzhong Art Museum, 2017

" The First 'Belt and Road' Africa-China Art Exchange Exhibition & Africa-China Culture Forum", National Library of China, 2018

"From Lausanne to Beijing", the 10th International Fiber Art Biennale Exhibition, Tsinghua University Art Museum, 2018

" Silk Road Art Charm, China, Europe, Female Art Exchange Exhibition", Prince Gong's Mansion, Beijing, 2018

" Different Voices", Care Narrative of Fiber Art, Joint Exhibition of 10 Artists from 10 Countries, Tsinghua University Academy of Fine Arts Museum, 2018

" The 13th National Art Exhibition", Shandong University of Art & Design, 2019

" The 5th International Art and Science Exhibition and Symposium", National Museum of China, 2019

2022"从洛桑到北京"国际纤维艺术双年展（云南·澜湄展年），云南昆明，2022

"唯物思维：首届国际当代材料艺术双年展"，清华大学、青岛国信集团，2022

《乐山水》，72△青年艺术家计划——"待到山花烂漫时"展览，江西景德镇，2022

获奖

《夜与昼》获"第十届全国美术展览"金奖，2004

《韵》获"第十一届全国美术展览"优秀奖，2009

《2014年亚太经合组织（APEC）会议领导人服装设计》获荣誉奖，2014

《传统苏绣工艺服装设计系列》入选"中国设计权利榜"，2018

《风声》获"第十三届全国美术展览"入围奖，2019

"夜与昼——李薇艺术作品展"获"北京国际设计周"月桂奖、年度设计师奖，2019

《紫气东来》获"北京2022年冬奥会和冬残奥会制服装备视觉外观设计"铜奖，2021

《无题》获"从洛桑到北京"第五届国际艺术双年展优秀奖，2008

《清·远·静》获"从洛桑到北京"第六届国际纤维艺术双年展金奖，2010

《夜与昼》获"艺术与科学国际作品展"优秀奖，2011

《Xiuniang·李薇高级定制时装发布会，梅赛德斯奔驰中国国际时装周》获第20届中国国际时装周十佳时装设计师，2014

" 2020 China and USA Technology & Innovation in Fiber Art Virtual Exhibition", Tsinghua University Academy of Arts & Design (Beijing, China), the University of the Arts (Philadelphia, USA), 2020

2021 "Integration of Art and Science, Quality Innovation for the Future", series theme exhibitions, Tsinghua Qingdao Academy of Art and Sciences, 2021

" Homomorphic, Contemporary Art Invitational Exhibition", Luohu Art Museum, Shenzhen, 2021

" Cloud Gathering: Annual Fashion Retrospective Exhibition 2011-2020", China National Silk Museum, 2021

"From Lausanne to Beijing", International Fiber Art Biennale Exhibition (Yunnan-Lanmei Exhibition Year), Kunming, Yunnan, 2022

" MATERIAL THINKING 1st International Contemporary Material Art Biennale", Tsinghua University, Qingdao Conson Development (Group) Co., Ltd., 2022

" Le Shan Shui", 72 Youth Artist Project "When Spring Comes" Exhibition, Jingdezhen, Jiangxi, 2022

Key Awards

" Night and Day" won the Gold Award at the 10th National Art Exhibition, 2004

" Rhyme" won the Excellence Award at the 11th National Art Exhibition, 2009

" 2014 Garment Design for APEC Economic Leaders" won the Honorary Award, 2014

" The Traditional Su Embroidery Clothing Design Series" was selected for the "China DESIGN POWER 100" List, 2018

" Sound of the Wind" won the Finalist Award at the 13th National Art Exhibition, 2019

" Night and Day - Li Wei Art Exhibition" won the Laurel Award and Designer of the Year Award at the Beijing International Design Week, 2019

" Purple Air from the East" won the Bronze Award for the visual appearance design of the Beijing 2022 Winter Olympics and Paralympics uniforms, 2021

" Untitled" won the Excellence Award at "From Lausanne to Beijing", the 5th International Fiber Art Biennale Exhibition, 2008

" Qing Yuan Jing" won the Gold Award at "From Lausanne to Beijing", the 6th International Fiber Art Biennale Exhibition, 2010

" Night and Day" won the Excellence Award at the International Art and Science Exhibition and Symposium, 2011

" Xiuniang, Li Wei Haute Couture Fashion Show, Mercedes Benz China International Fashion Week" won the Top 10 Fashion Designer at the 20th China International Fashion Week, 2014

作品收藏

《凤袍》法国高级时装设计师RANADA收藏，2003

《夜与昼》中国美术馆收藏，2004

《清·远·静》河南美术馆收藏，2010

《夜与昼》华南农业大学设计学院美术馆收藏，2011

《新中式·旗袍》中国丝绸博物馆收藏，2012

《新中装》被英国维多利亚与阿尔伯特博物馆永久收藏并作为常设展展出，2019

《基因》被中国丝绸博物馆收藏，2021

《青绿山水》被中国丝绸博物馆收藏，2021

《宋韵数字时装》被中国丝绸博物馆收藏，2023

作品拍卖

《述说》参加中国嘉德秋季"灵感–艺术设计专场"被拍卖，2012

《夜与昼》参加中国嘉德秋季"灵感–艺术设计专场"被拍卖，2013

《清·远·静》参加中国嘉德秋季"灵感–艺术设计专场"被拍卖，2014

Artwork Collections

" Fengpao" collected by French haute couture designer RANADA, 2003

" Night and Day" collected by the National Art Museum of China, 2004

" Qing Yuan Jing" collected by Henan Art Museum, 2010

" Night and Day" collected by the Art Museum of School of Art and Design, South China Agricultural University, 2011

" New Chinese-style Cheongsam" collected by the China National Silk Museum, 2012

" New Chinese-style" permanently collected by the Victoria and Albert Museum in the UK and exhibited as a permanent exhibition, 2019

" Gene" collected by the China National Silk Museum, 2021

" Green Landscape" collected by the China National Silk Museum, 2021

" Digital Fashion of Song Rhyme" collected by the China Silk Museum, 2023.

Artwork Auctions

" Narration" participated in the "Inspiration - Art Design Session" of China Guardian, Hong Kong, 2012 Autumn Auctions

" Night and Day" participated in the "Inspiration - Art Design Session" of China Guardian, Hong Kong, 2013 Autumn Auctions

" Qing Yuan Jing" participated in the "Inspiration - Art Design Session" of China Guardian, Hong Kong, 2014 Autumn Auctions

内 容 提 要

本书是著名艺术家、教育家、服装设计师清华大学美术学院教授李薇的艺术作品集，作品包括《夜与昼》《七彩云南》《青绿山水》《新中式》《蓝之韵》《红色记忆》《紫气东来》《锦瑟华年》《荷塘月色》《雅韵》《花梦敦煌》《空与影》《述说》《清·远·静》《行云流水》《风声》《夜与昼·丝网版画》《山峦》《枯山水》《大漠孤烟》《乐山水》。

本书较为全面地收录了李薇教授四十余年创作的当代艺术、时装艺术、纤维艺术作品，是其艺术成就的展现和总结。

图书在版编目（CIP）数据

衣者·天地形 / 李薇著 . -- 北京：中国纺织出版社有限公司，2023.10
ISBN 978-7-5229-1046-8

Ⅰ. ①衣… Ⅱ. ①李… Ⅲ. ①服装设计－作品集－中国－现代 ②纤维－编织－工艺美术－作品集－中国－现代 Ⅳ. ① TS941.28 ② J523.4

中国国家版本馆CIP数据核字（2023）第175948号

责任编辑：亢莹莹　　特约编辑：徐屹然
责任校对：高　涵　　责任印制：王艳丽

图书设计：杨　晋　　Sasha Yang　　曹一齐
执行团队：毕　然　　王　玲　　刘　魋

中国纺织出版社有限公司出版发行
地址：北京市朝阳区百子湾东里A407号楼　邮政编码：100124
销售电话：010—67004422　传真：010—87155801
http://www.c-textilep.com
中国纺织出版社天猫旗舰店
官方微博 http://weibo.com/2119887771
北京华联印刷有限公司印刷　各地新华书店经销
2023年10月第1版第1次印刷
开本：787×1092　1/8　印张：27.5
字数：190千字　定价：380.00元

凡购本书，如有缺页、倒页、脱页，由本社图书营销中心调换